Human Body

Human Body

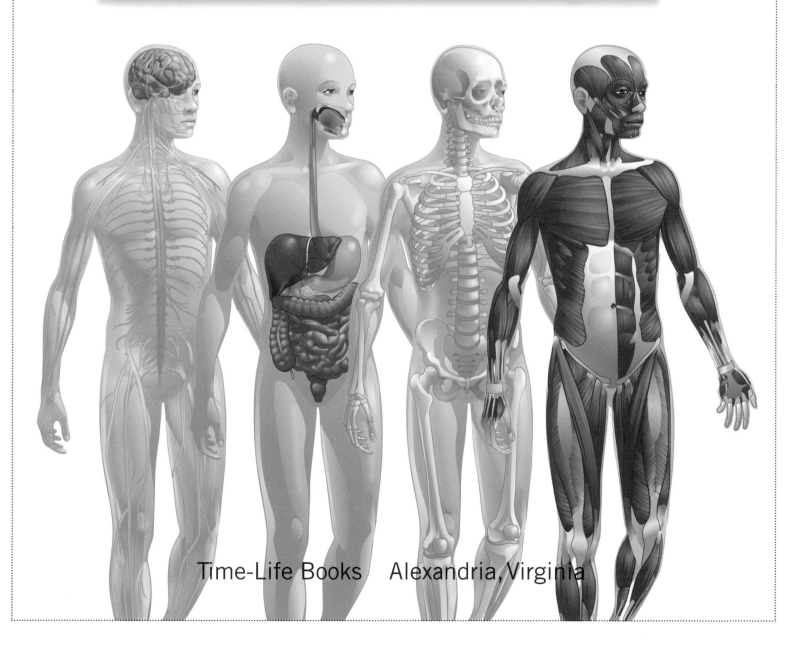

Time-Life Books Alexandria, Virginia

Table of Contents

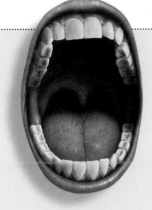

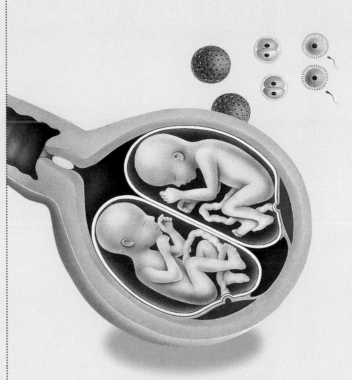

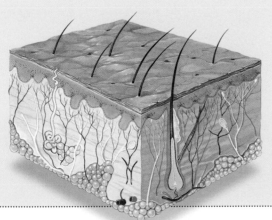

The Human Body

A human being is a complicated collection of bones and muscles, blood and organs, thoughts and feelings, and knowledge. Controlling everything is the brain, with more intelligence than that of any other animal. What's really amazing is how all this complexity starts as a single cell—a fertilized egg. But this cell is not so simple. Packed into its center—the nucleus—is everything needed to make a human body. Once the body is made, it takes years to grow and develop into a finely tuned flesh-and-blood machine that can reason and move anywhere its brain tells it. Whether you are walking or tumbling head over heels on a gym mat, your brain keeps you balanced and positioned just right. It analyzes a steady stream of information coming from your eyes, mouth, nose, ears, and skin. Messages from your brain then zoom along nerve pathways to tell your muscles what to do. Your heart beats. You breathe. Your eyes blink. You swallow. You do a perfect backflip.

Skeleton The Body's Framework

Your skeleton is a lot more than a collection of bones. Like the steel girders of a tall building or the trunk and branches of a tree, it forms the framework for your body. Your skeleton gives you a shape, supports your body's weight, and anchors your **muscles** so you can move. It is strong enough to protect delicate internal **organs,** such as your heart, lungs, and brain, and yet light enough to allow you to jump, climb, and swim. The thighbones alone can absorb up to 58,000 kg of pressure per sq cm (20,000 lb./sq in.).

From the skull to the toes, the adult human skeleton contains 206 bones and weighs about 9 kg (20 lb.). A baby's skeleton contains 350 bones, but these are soft and fuse together as the baby grows. The skeleton is not completely hardened and fused together until a person is about 30 years old.

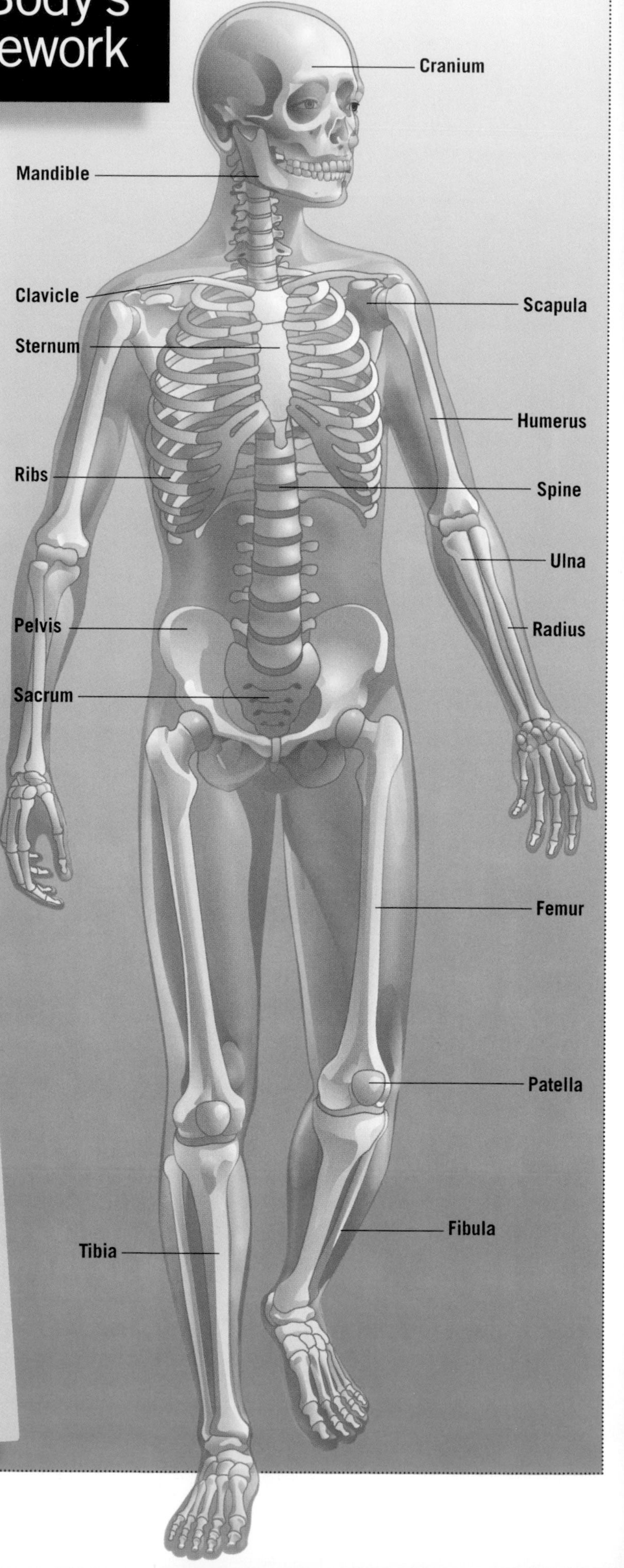

Cranium

Mandible

Clavicle

Scapula

Sternum

Ribs

Humerus

Spine

Pelvis

Ulna

Sacrum

Radius

Femur

Patella

Fibula

Tibia

Would **You** **Believe?**

How Quickly Bones Form

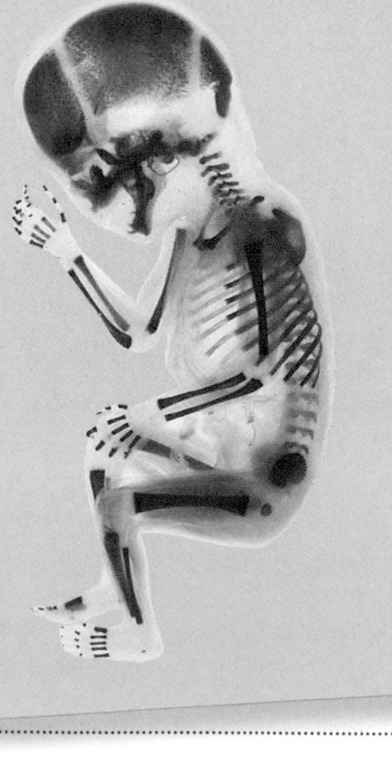

Can you guess how old this baby is? It hasn't been born yet! This unborn baby, called a **fetus,** is only 12 weeks old. It's tiny—it weighs 3 g (0.10 oz.) and is 9 cm (3.5 in.) long—but it already has many of the bones that will make up its adult skeleton. The darker parts are the bones, and the lighter parts are a soft substance called **cartilage** that will later become bone. Some cartilage will remain, however. The tip of your nose and your ears are bendable because they are made of cartilage.

Largest and Smallest Bones

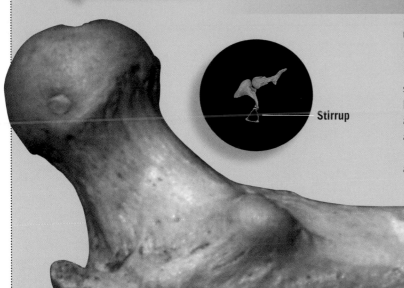

Stirrup

The longest bone in the skeleton is the femur, or thighbone, which connects the hip and the knee. It is the heaviest and strongest bone in the body. In most people the femur is about a quarter of their total height, usually about 37 to 45 cm (15 to 18 in.) long. The three smallest bones are found in the ear. The tiniest of these is the stirrup. It measures only 0.5 cm (0.25 in.) long. The bones here are shown at their actual size.

Bone Size

How tall you are depends on the length of your bones. Tall people have long bones; the bones of short people are shorter.

Your growth is controlled mainly by chemicals in your blood called **hormones.** Sometimes these hormones don't work the way that they're supposed to, making bones grow too quickly or too slowly during childhood. This can make a person unusually tall or unusually short.

The tallest person on record, an American named Robert Wadlow, measured 2.7 m (8 ft. 11 in.) and weighed 198 kg (439 lb.). One of the shortest people, and certainly the lightest, was a Mexican woman named Lucia Zarate. She was about 66 cm (26 in.) tall and weighed 2.1 kg (4.7 lb.) when she was 17.

Robert Wadlow towers above his regular-size family, whereas tiny Lucia Zarate perches like a doll on a tabletop.

Strange But TRUE!

Drink Your Vitamins

If you don't get enough **calcium** or vitamin D, your bones might become as soft and twisted as the skeleton shown at right. The skeleton's owner suffered from rickets, a disease in which the bones do not become hard. Rickets used to be a common problem for children, and it still is a problem in those parts of the world where milk and food with vitamin D are not available.

Skull A Box for the Brain

I f you feel your own skull with your fingers, it feels like one solid bone. It's not. The skull is actually made up of 28 bones whose jagged edges lock together like the pieces of a jigsaw puzzle. The places where these bones join are called sutures. Although they look like cracks (such as on the skull at right), they actually make the skull stronger. If you hit your head, the sutures absorb the shock wave from the blow.

Scientists divide the skull into three main parts: the **cranium** (the round part that surrounds your brain), the face, and the mandible (lower jaw). The cranium is made of eight bony plates that form a smooth, protective bowl. Nine bones make up the face. Some of these bones have a tiny hole in them called a foramen that allows **nerves** or blood vessels to pass through. Inside the ears are three more bones that help you hear. The mandible is the only bone in the skull that can move, allowing you to open and close your mouth.

Suture

Frontal Bone

Parietal Bone

Foramen

Temporal Bone

Nasal Bones

Zygomatic Bone

Foramen

Maxilla

Mandible

Foramen

Let's Compare

Growing Skulls

Looking down on the heads below, you can see how a skull changes as it grows. An infant's skull has large soft spots, called fontanels, between its bony plates. These allow the skull to be squeezed slightly as the baby is born and make room for the brain to grow. They harden into bone by the time a baby is two years old and are fully fused together in an adult. But the joints, or sutures, remain.

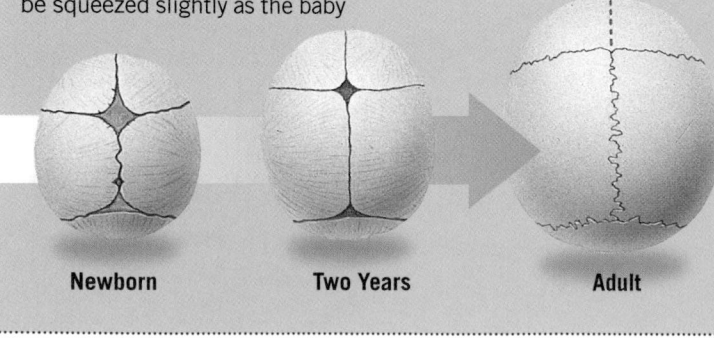

Newborn   **Two Years**   **Adult**

An Architect's Dream

What is curved, rigid, and very strong but also very light? Your skull! One Italian architect used these qualities to design a roof for a sports arena in Rome (below). This technique, called thin shell construction, uses a thin layer of concrete that is made of small pieces and curved like a cranium. Like the skull, it is exceptionally strong for its weight.

Phineas Gage

In 1848, when Phineas Gage was 25 years old and working on a railroad in Vermont, he was preparing a gunpowder charge that exploded before he was ready. The explosion drove a 1-m (3-ft.)-long iron bar completely through his skull. Although the bar passed under the eye socket and fractured some bones, Gage remained conscious and later recovered from his injury, except that he lost sight in his left eye and had a dented skull.

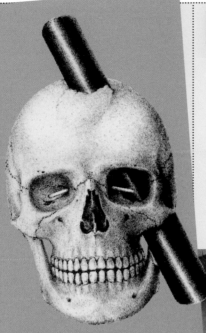

A medical illustration shows where the bar passed through Gage's skull before it exited through the top of his head.

Skulls in Culture

Skull Candy

Would you eat candy shaped like a skull? You might if you lived in Mexico, where people celebrate the Day of the Dead on November 2. On this day they honor members of their families who have died.

Ancestral Spirit

Skulls have a symbolic power around the world. In Tibet, monks drink from cups made from skulls. On the island of Java, warriors like the man in this photograph sometimes sleep on a skull to become closer to the ancestor who originally owned it.

Sports: Hazardous to Your Head?

Any sport that involves repeated blows to the head can injure your brain or the bones in your skull, particularly the jaw. That's why it is important to wear protection such as helmets or jaw guards. Boxing is the most hazardous. Repeated blows make boxers "punch drunk," a condition that causes slurred speech and loss of balance—very much like a person who has had too much alcohol. But there are also risks in other high-contact sports such as soccer and Rugby.

Would **You** Believe?

Stone Age Surgery

Modern people are not the only ones to practice brain surgery. Even cave dwellers did it, judging from the prehistoric skulls, marked by large holes, that archaeologists have found. No one knows why early people tried this surgery. Perhaps they wanted to help people suffering from skull fractures or terrible headaches called migraines. At first, ancient people used stone tools to open skulls. Later, they worked with metal tools, like this bronze blade from Peru.

Spine Ribs, and Hips

Three parts of the skeleton shape and protect the human torso: the spine, the ribs, and the hips. Your spine, or backbone, is strong and flexible, so you can twist and bend side to side, forward, and back. It is also elastic, absorbing jolts as you walk or run.

Your backbone is not really one bone but 33 linked together. These spinal bones are called vertebrae. The top two vertebrae are shaped differently from the rest. The first vertebra, called the atlas after the Greek god who carried the earth on his shoulders, supports the head. The second vertebra is called the axis. The joint between these two vertebrae allows you to nod or shake your head.

Attached to the middle section of your spine are your ribs, which protect the heart and lungs. Your hips, which make walking possible, connect to the bottom of your spine.

Anatomy of a Vertebra

This closeup of a typical vertebra shows the many uses of the spine. The thick, round part is called the body; it supports your weight. The spiky "wings" sticking out the sides attach to the **muscles.**

A bony hole at the center protects the **spinal cord,** a bundle of **nerves** that runs through the vertebrae. The spinal cord and the spinal nerves coming off it carry messages between your brain and the rest of your body.

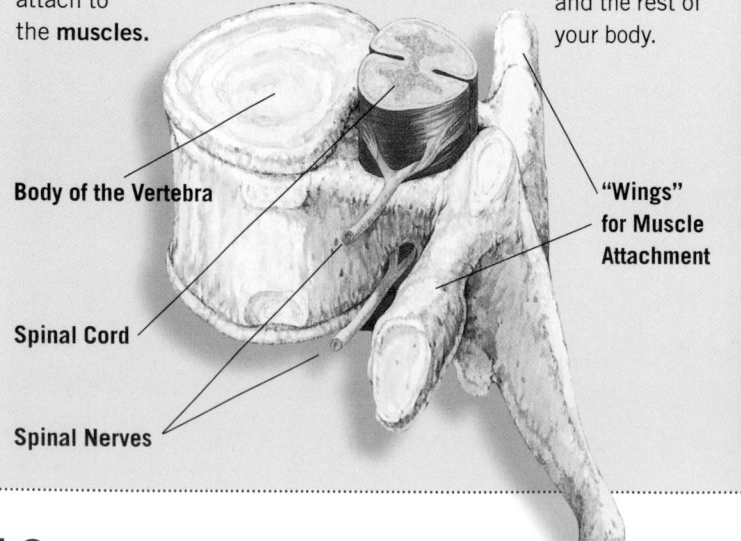

Body of the Vertebra

Spinal Cord

Spinal Nerves

"Wings" for Muscle Attachment

Even when you stand up straight, your backbone bends. The slight S curve helps balance the load of your head and torso, the middle section of your body. Between each vertebra is a cushiony disk of **cartilage** to keep the bones from rubbing against each other. Because these disks flatten out from the weight of your body during the day, most people go to bed shorter than when they wake up! At the bottom of your spine is the coccyx, or tailbone—one of the few useless bones in your body.

The Spine

Atlas

Axis

7 Cervical (Neck) Vertebrae

12 Thoracic (Upper Back) Vertebrae

Disk

5 Lumbar (Lower Back) Vertebrae

5 Fused Sacral (Pelvic) Vertebrae

4 Fused Tail Vertebrae (Coccyx)

Hip Shape

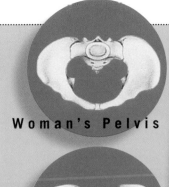

Woman's Pelvis

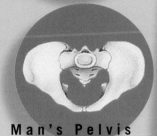

Man's Pelvis

The place where the two hip-bones meet the bottom of the spine is called the **pelvis.** As you can see, a woman's pelvis is shaped differently from a man's. To allow a baby to pass between these bones, a woman's pelvis is wider and more flexible.

Oh, My Back!

Imagine being told by your doctor that the only cure for your backache was to be tied to a ladder, hoisted up high, and then dropped! Well, that's what is being done to the poor back sufferer in this 10th-century illustration. In those days people thought if you gave the back a jolt, any displaced bones would fall into place. What a jolt!

Would You Believe?

A long neck is considered beautiful in many parts of the world, but few people would go this far to get one! In Myanmar (Burma), the women of the Padaung tribe stretch out their vertebrae and lengthen their neck by wearing metal rings. Beginning with a few rings at age five, girls gradually add more rings as they grow older. They wear them even when they sleep!

Cage for the Heart and Lungs

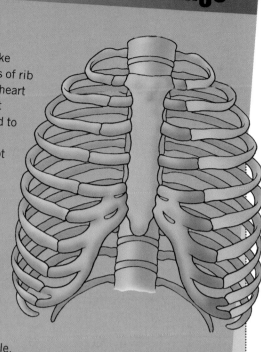

Circling your chest like curved bars are 12 pairs of rib bones that protect your heart and lungs. They connect to your spine in back and to your flat breastbone, or **sternum,** in front—except for the bottom two pairs, which are called floating ribs because they are not attached at the front. The rib cage is strong and flexible. Muscles between your ribs allow the cage to expand—get bigger—so air can fill your lungs when you breathe, and then contract—get smaller—to force air out when you exhale.

Computer Smarts

Do you spend a lot of time sitting in front of a computer? If so, you could be hurting your back and wrists. But you don't have to. If you take frequent breaks and sit like the girl at right, you will be doing your body a big favor. She's sitting up straight, with her back supported by a good chair. Her elbows and knees are bent at a 90-degree angle, and her wrists are straight. Bent wrists put a lot of strain on nerves, muscles, and **tendons,** and can lead to a painful condition called carpal tunnel syndrome.

The Amazing Hand

With your hands you can do an amazing variety of things: You can pound a nail, stroke a kitten, tie a shoelace, push a button, hang from a chin-up bar, play a piano, or pick up a pin. If you know sign language, your hands can even talk! All these movements are possible because of the 54 bones in your hands and the **muscles** and **ligaments** that connect them.

The bones of the hand are divided into three types: the carpals, metacarpals, and phalanges. The carpals are the small, pebble-shaped bones in your wrist. Branching out from the wrist in a fan shape are the five long metacarpal bones, which you can feel on the back of your hand. Joining the metacarpals at the knuckles are the phalanges. Each finger has three phalanges, except the thumb, which has two.

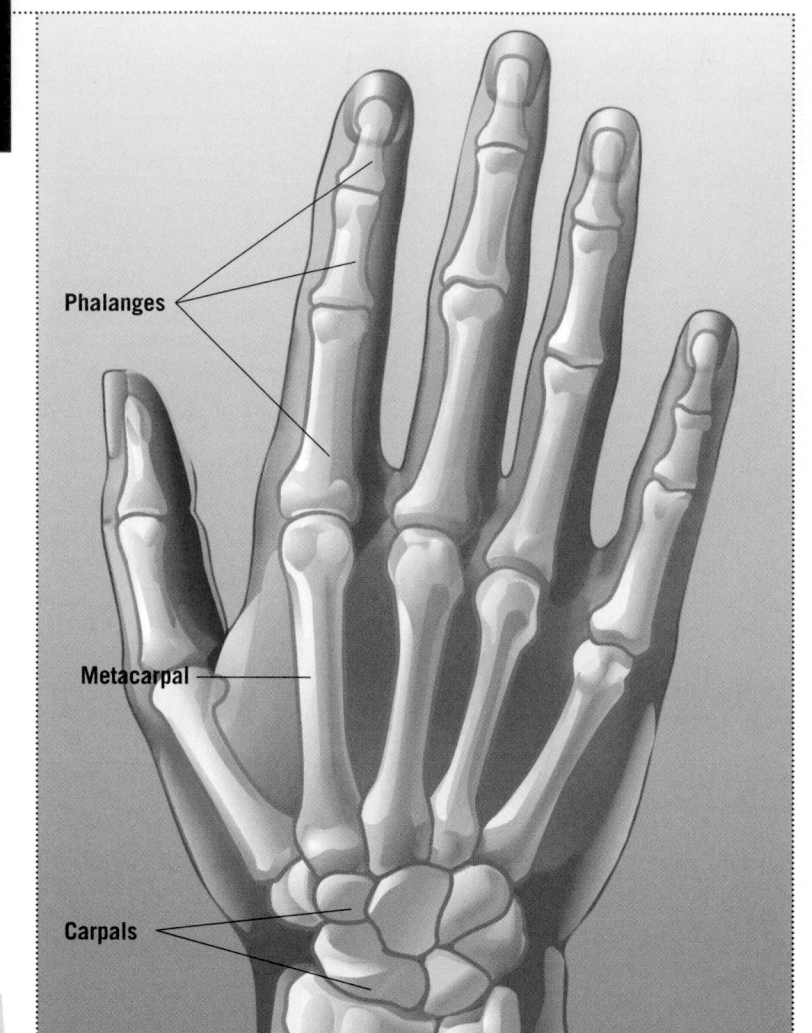

Phalanges

Metacarpal

Carpals

Famous 1 FIRSTS

X-Rays

In 1895 Wilhelm Roentgen, a German scientist, discovered that a certain kind of radiation could pass through skin, paper, wood, and other materials. He didn't know what to name his discovery, so he called it x-radiation, with the x meaning "unknown." The x-ray at right is one of the first: It is Mrs. Roentgen's left hand. The object on her third finger is her wedding ring. Today, the x-ray is a powerful tool for the diagnosis of disease.

Bone Growth

As you grow, so do the bones in your hands, as you can see in the x-ray photographs below. Although the finger bones look disconnected in the infant's hand, they are actually attached with soft **cartilage,** which will eventually harden into bone. The bones in the teenager's hand still have room to grow. By the time the teenager becomes an adult, the hand bones stop growing and the cartilage disappears.

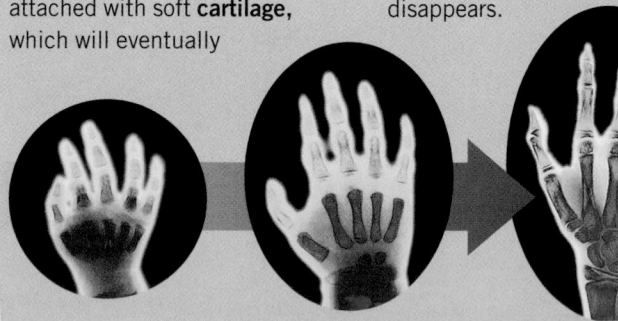

Infant   Teenager   Adult

The Versatile Hand

The human hand, with its opposable thumb, can grip objects in a surprising variety of ways. You can see the main grips in the photographs below.

Precision Grip

The precision grip, with the thumb opposite the index, or pointer, finger, is used in writing with a pen or pencil, holding chopsticks, or drawing a bow across the strings of a cello.

Power Grip

Holding a baseball bat or tennis racket in the power grip, with the muscles of your hand wrapped firmly around the object, gives the most power to your swing.

Hook Grip

You use a hook grip, with your fingers curved in the shape of a hook, to carry a suitcase by the handle or to grab on to a ledge when rock climbing.

Oblique Grip

With the sides of your thumb and forefinger together in the oblique grip, you can hold and thread a needle.

Saving Fingers

Although it sounds impossible, doctors have been able to reattach fingers after they have been cut off *(right)*. By carefully reconnecting the damaged **nerves** and blood vessels during an operation, surgeons make it possible for the rest of the hand to heal itself. Even a bone that has been cut will grow together again as long as it has a blood supply.

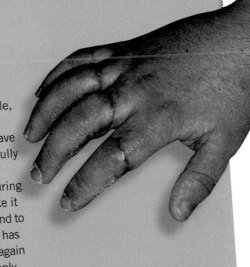

Let's **Compare**

Human versus Chimp Hand

Chimpanzees have hands very similar to ours, but with one big difference: Their thumb bones are not as long as ours. Because of our long thumb bones, our thumbs can touch the tips of each of the other fingers. This kind of thumb is called an "opposable" thumb, because it works against, or "opposes," the fingers, making it possible for us to pick up small objects and hold a pencil or a dart in a precision grip. No other creature has an opposable thumb.

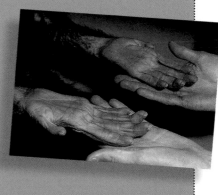

Inside Bones

Bones in a museum are dead, but inside your body, bones are very much alive. They have living **nerves** and blood vessels running through them. They grow, change, and heal themselves. Bones are made of **minerals** and **collagen.** Minerals make bones hard; collagen makes bones slightly elastic, so they don't easily snap. Like a loaf of crusty bread, bones are hard on the outside and full of holes on the inside. The tough outer layer is called compact bone. The inside layer is called spongy bone, but it isn't soft. It is almost as strong as compact bone but much lighter because it is full of small, empty spaces. In the middle of your large bones is **bone marrow,** which is hard at work creating new blood for the rest of your body. Covering bones all over, except where they join other bones, is a tough, thin "skin" called periosteum, which contains blood vessels and nerves. It helps bones heal when they break.

Keeping Bones Healthy

These pictures show what bone looks like through a microscope. The one on top is a healthy bone. The picture below shows a bone that has been eaten away by a disease called osteoporosis, which affects many older people, particularly women. Osteoporosis happens when bones lose **calcium** and become brittle and more likely to break. To prevent this disease, eat foods rich in calcium and in vitamin D, which helps the body absorb calcium. You also need to exercise, because the more you use your bones, the stronger they get.

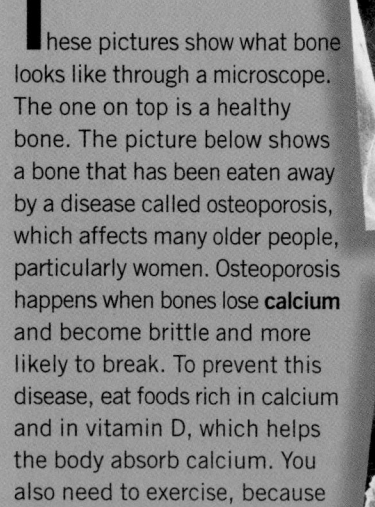

Ulnar Nerve

What's the Funny Bone?

Sometimes, if you hit your elbow in just the wrong spot, a sharp, fizzing pain will run up your arm and last for several seconds. This is called "hitting your funny bone," but there's nothing funny about it. In fact, there really isn't a bone called the "funny bone." What you are hitting is a nerve called the ulnar nerve that runs down your arm. When you hit it, the nerve quickly sends a painful message to your brain that makes you say, "Ouch!"

Bone Architecture

Like the arching iron grid-work of the Eiffel Tower in Paris, bones are curved and reinforced for strength. The secret of bone's great strength lies inside. If you slice through a femur *(left)*, the strongest bone in the body, you will see many tiny curved ridges of bone. Extra ridges form naturally where bones bear the most weight. This makes bones stronger where they need to be. Imagine the pressure put on the leg bones when Olympian Jackie Joyner-Kersee *(far left)* lands a successful long jump.

How Bones Heal

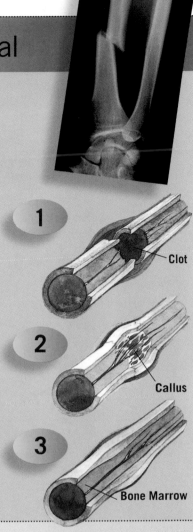

Broken bones, such as the shinbone at right, are called fractures. They begin healing themselves as soon as they break. First, blood rushes to the scene and hardens into a clot around the break (1). Then the ends of the bones become soft as minerals start to seep out of them into the gap. New bone called callus starts to grow from the broken ends (2). Finally, the callus starts to harden into true bone, and the bone can be used again (3). Healed bones are as strong as—and sometimes even stronger than—bones that have never been broken.

1

Clot

2

Callus

3

Bone Marrow

Would **You** Believe?

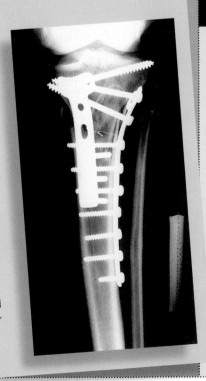

Bones, although strong, can break, sometimes in many places. The x-ray at right shows the leg bone of a man who crashed his motorcycle at 257 km/h (160 mph). Both his legs were shattered in so many spots that doctors had to use 26 metal screws and plates to put them back together. A bone doctor, or orthopedist, will usually set a broken bone by lining it up and wrapping it in a plaster cast for a few weeks or months. The bone then heals itself.

People Crazy Sally

People with broken bones in 18th-century England did not go to doctors. They went to bonesetters who traveled the country fixing bones. One of these was Mrs. Sarah Mapp, also called Crazy Sally. Crazy Sally was so strong she could shove a dislocated shoulder back in place and straighten a man's leg that had been crooked for 20 years. Some people made fun of her appearance *(cartoon at right)*, but she was good at what she did.

What Are Joints?

When you kick a ball, bend your finger, stand on tiptoe, or nod your head, you are using your **joints.** Joints are where your bones connect to each other. They make it possible for you to move.

Some joints move a lot, some move a little, and some don't move at all. For instance, your hip joint moves all around. The joints of your spine move just a little. The suture joints in your skull don't move.

Tough, ropy **ligaments** hold together the bones in your joints. Ligaments allow your joints to move in the right direction and make sure they don't move too far in the wrong direction. Some people are called double jointed because they can pull their fingers back very far or twist their spines into weird shapes, like the circus contortionists in the photograph. No one really has double joints, though. People who can move this way can do it because their ligaments can stretch much farther than usual.

The Amazing Knee Joint

The knee joint works in one direction only, opening and closing like a hinge. But even this simple movement requires a complicated joint *(drawing at right)*. Outer and inner ligaments hold the thigh and shin leg bones securely, and pads of **cartilage** cushion the inside of the joint. A special liquid called synovial fluid keeps the joint moving smoothly like a well-oiled machine. **Tendons** hold the knee-cap, or patella, which acts like a shield to protect the joint.

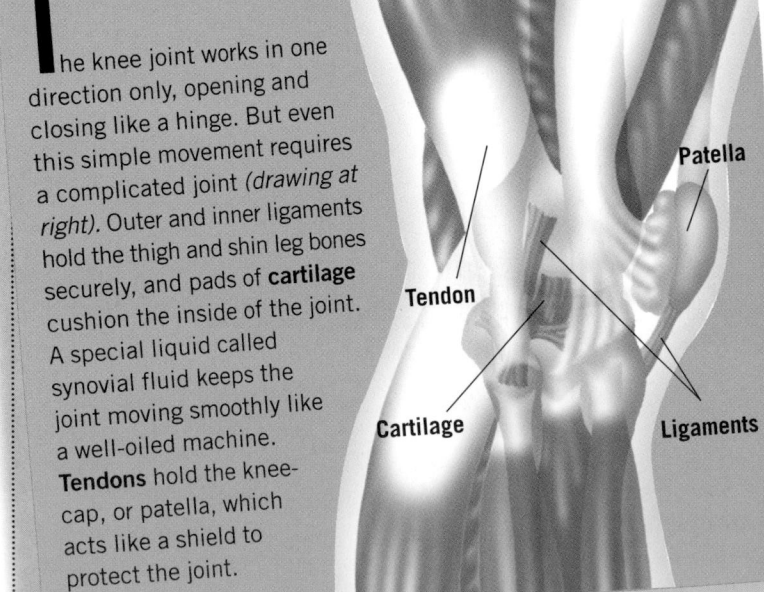

Patella

Tendon

Cartilage

Ligaments

Different Types of Joints

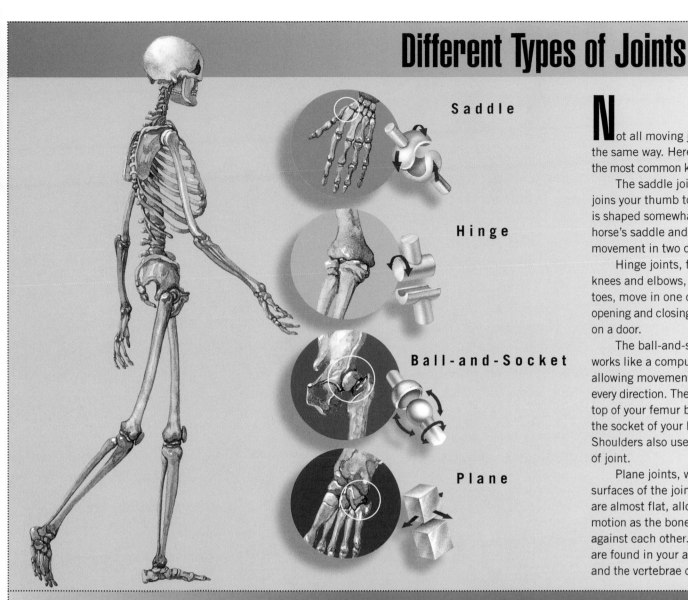

Saddle

Hinge

Ball-and-Socket

Plane

Not all moving joints work the same way. Here are some of the most common kinds of joints:

The saddle joint, which joins your thumb to your wrist, is shaped somewhat like a horse's saddle and allows movement in two directions.

Hinge joints, found in your knees and elbows, fingers and toes, move in one direction, opening and closing like a hinge on a door.

The ball-and-socket joint works like a computer joystick, allowing movement in almost every direction. The ball-shaped top of your femur bone fits the socket of your hip like this. Shoulders also use this type of joint.

Plane joints, where the surfaces of the joining bones are almost flat, allow a gliding motion as the bones slide against each other. These joints are found in your ankle, wrist, and the vertebrae of your spine.

Artificial Joints

Human joints take a huge amount of abuse. They continue to work smoothly long after the moving parts of many machines would wear out. But sometimes joints wear out, too, particularly when they are affected by a crippling disease called arthritis. When someone has arthritis, the joints get stiff and swell painfully. Eventually the joints may break down. If the condition gets too bad, doctors replace the natural joints with artificial ones made of steel and plastic. Surgeons remove the diseased or worn-out joint and insert the replacement, which works nearly as well as the original. People whose hips or knees were crippled by arthritis can walk again with artificial joints. Knees and hips aren't the only joints replaced. There are artificial joints for shoulders, elbows, wrists, and ankles, too.

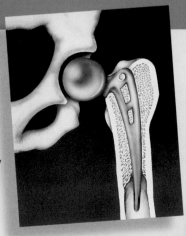

The diagram above shows an artificial hip joint (*blue*) in place.

Is "Cracking" Bad for Joints?

When people "crack" their knuckles or do deep knee bends, you often hear a sharp snapping. It sounds painful, and to some it is very annoying. Doctors don't really know what causes the sound but say there is no evidence that cracking joints is harmful to them.

What Are Muscles?

Although **joints** allow bones to move in a certain direction, **muscles** do the actual work. Muscles —more than 600 of them—cover your entire skeleton, as you can see in the illustrations at right. But other muscles inside your body allow movements you probably don't even think about: your heartbeat, the in and out of your breathing, and the rippling motion of your **esophagus** carrying food.

Muscles are made of bundles of very thin fibers, like strings. Some attach directly to bone, whereas others are attached to your bones by **tendons,** which are like tough, strong ropes. The largest muscles are in your legs, buttocks, and arms, because these muscles move your heaviest bones. All together your muscles make up about one-third of your total weight.

Types of Muscle

The muscles in your body can be divided into three types: cardiac, smooth, and skeletal. Cardiac muscle is found only in the heart. Smooth muscle (which looks smooth under the microscope) controls movement in the other internal organs, keeping the blood moving in your **arteries,** moving food through your **digestive system,** and getting rid of wastes. Both cardiac and smooth muscle are called involuntary, because you can't control them.

Skeletal muscle—sometimes called striated or striped because of its appearance—is attached to your bones. You can control it when you want to move your finger or lift your leg.

Cardiac

Smooth

Skeletal

Frontalis
moves eyebrows and forehead

Orbicularis Oris
"kissing muscle"

Orbicularis Oculi
closes eyelids

Biceps
bends elbow

External Oblique
twists and bends body

Quadriceps Femoris
straightens knee

Sartorius
(longest muscle)
bends knee and twists leg

Extensor Digitorum Longum
bends toes upward

Tibialis Anterior
bends ankle, turning sole inward

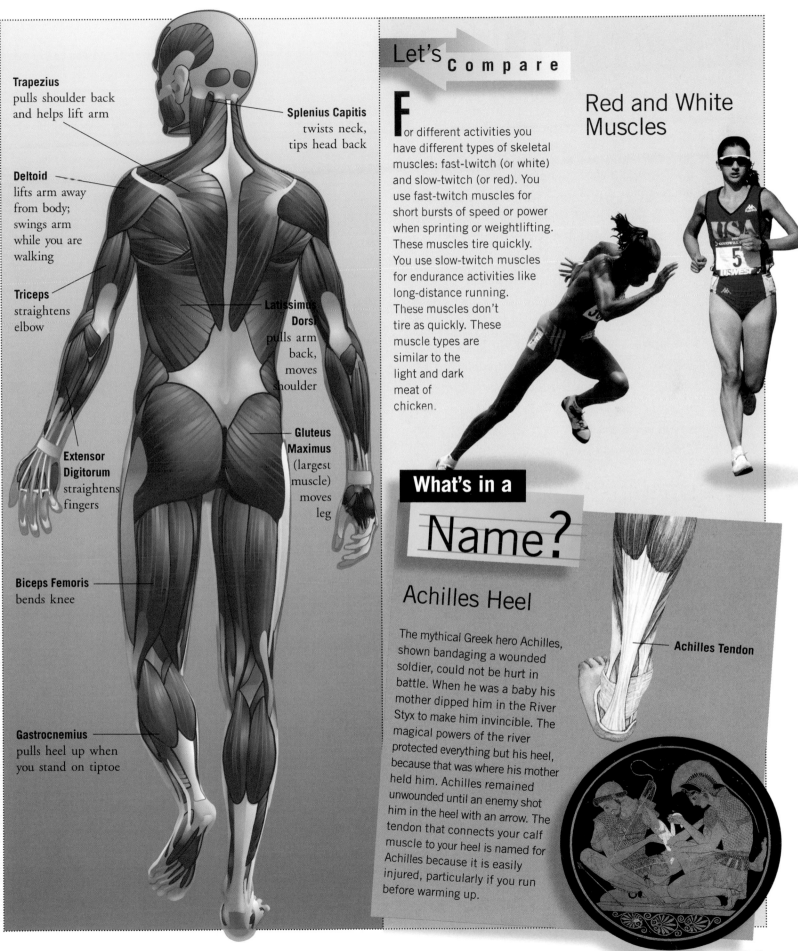

Trapezius
pulls shoulder back and helps lift arm

Splenius Capitis
twists neck, tips head back

Deltoid
lifts arm away from body; swings arm while you are walking

Triceps
straightens elbow

Latissimus Dorsi
pulls arm back, moves shoulder

Extensor Digitorum
straightens fingers

Gluteus Maximus
(largest muscle) moves leg

Biceps Femoris
bends knee

Gastrocnemius
pulls heel up when you stand on tiptoe

Red and White Muscles

For different activities you have different types of skeletal muscles: fast-twitch (or white) and slow-twitch (or red). You use fast-twitch muscles for short bursts of speed or power when sprinting or weightlifting. These muscles tire quickly. You use slow-twitch muscles for endurance activities like long-distance running. These muscles don't tire as quickly. These muscle types are similar to the light and dark meat of chicken.

What's in a Name?

Achilles Heel

The mythical Greek hero Achilles, shown bandaging a wounded soldier, could not be hurt in battle. When he was a baby his mother dipped him in the River Styx to make him invincible. The magical powers of the river protected everything but his heel, because that was where his mother held him. Achilles remained unwounded until an enemy shot him in the heel with an arrow. The tendon that connects your calf muscle to your heel is named for Achilles because it is easily injured, particularly if you run before warming up.

Achilles Tendon

How Muscles Work

Muscles move by contracting—that is, by getting shorter. When cardiac **muscle** contracts, it pumps blood. When smooth muscle contracts, it squeezes food through your **digestive system.** When skeletal muscle contracts, it pulls the bone it is connected to, lifting it.

Skeletal muscles contract when they get a signal from the brain. To make one movement, such as lifting a glass of water, the brain sends signals to many muscles. Even when you're standing still, dozens of your muscles must contract just to keep you from falling over.

Muscles can only pull—they cannot push. Most skeletal muscles work in opposing pairs: When one muscle contracts, its partner on the other side of the bone relaxes. Some exceptions to that rule are the eyelid muscles for blinking and the **diaphragm** muscle for breathing, which don't have partners.

Muscle Teams

Muscles, **ligaments, tendons,** and bones work together as a team to move the body. The shoulder and arm shown in the drawing at right represent this teamwork well. The arm bones are moved by skeletal muscles, which are attached to the bone by tendons. Ligaments surrounding the elbow hold the bones together in the **joint.** But nothing moves until the brain tells the muscles to move. The brain sends messages to the muscles through **nerves.** When you look at muscle **tissue** through a microscope (right), you can see nerves (green) attached to the muscle tissue (pink).

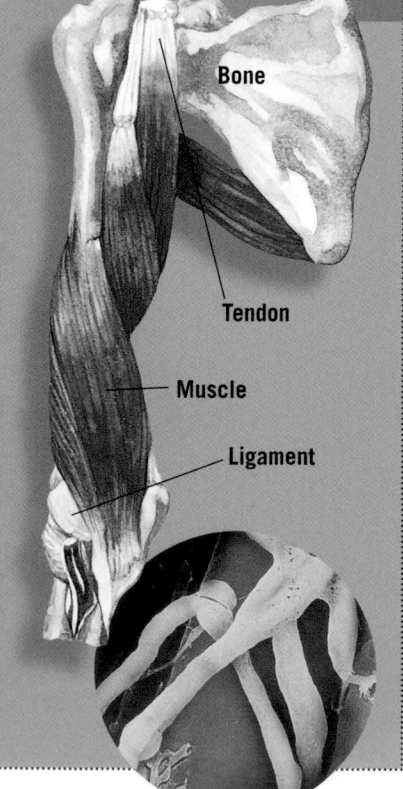

Bone

Tendon

Muscle

Ligament

How Muscles Pull

Contracting

When you bend your elbow, the biceps muscle of your upper arm contracts and bulges, pulling the forearm bones up. The opposing triceps muscle relaxes.

Biceps

Triceps

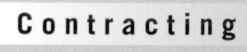

Extending

When you extend your arm, the muscle pair reverses its actions: This time, the triceps muscle contracts, pulling the forearm bones down, while the biceps muscle relaxes.

Like the strong cables that lift the arms of these big cranes, your muscles contract and pull on your bones to lift them.

How Strong?

Jaw Muscles

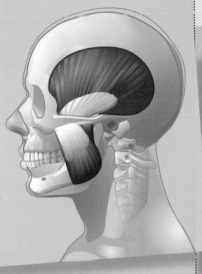

Some of the strongest muscles in your body are the powerful masseter muscles attached to your jawbones. You can feel these muscles in your cheeks when you clench your teeth. Their job is to open and shut your mouth firmly so your teeth can bite off chunks of food. These muscles can exert up to 90 kg (200 lb.) of pressure. Did you know that a human bites down with the same force as a shark? Scientists have proved this with an instrument called a gnathodynamometer.

Does a Body-builder Have More Muscles?

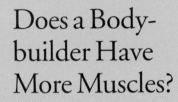

Bodybuilders like the ones shown here don't have more muscles, just bigger and more defined ones. They build these muscles by exercising regularly with weights. The more you use a muscle, the stronger and bigger it grows. Exercising a specific muscle makes the heart send it more blood and nutrients, which causes the muscle to grow more muscle fibers.

Communicating with Muscles

You use the muscles of your mouth and throat to speak, but sometimes you use other muscles to communicate nonverbally—without words. Here are some of the ways our muscles help us communicate.

Sign Language

Deaf people often use sign language to communicate with each other. Hearing people sometimes learn sign language, too. Each letter of the alphabet has its own finger arrangement, or sign, as do many words. Babies can learn to make some signs before they are able to speak.

Body Language

When you shake your head or shrug your shoulders, you are using body language to communicate. Sometimes we do this without realizing it, like the man on the camel who is gesturing with his hand as he talks on the telephone.

Facial Expression

Did you know it takes about 14 tiny facial muscles to make a smile? We show emotions in our faces: smiling to show joy, frowning to show anger, raising an eyebrow to show doubt. Watching someone's face when he or she talks will let you know how that person is feeling.

Why Do We Exercise?

The more you exercise your bones and **muscles,** the healthier and stronger they become. Without exercise, bones and muscles grow weak, and then they are more easily injured.

There are two main types of exercise: aerobic, which uses oxygen for energy, and anaerobic, which doesn't. Aerobic exercises strengthen your heart by making it work harder, and force your lungs to pull in more oxygen from the air for your muscles to use. Exercises that make you breathe hard, like swimming and jogging, are aerobic exercises. Anaerobic exercises, like sprinting or weightlifting, require a quick burst of energy. The lungs can't provide oxygen that fast, so anaerobic exercises use energy already stored in muscles. This builds up muscles by making them do more work.

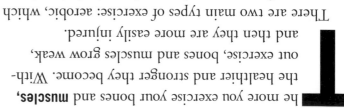

Heart Aerobics

Aerobic exercise, such as jumping rope, strengthens the most important muscle in your body—your heart. Muscles need oxygen to work, and they get it from the blood that the heart pumps to them. Oxygen enters the blood when your lungs breathe air. The harder your muscles work, the more oxygen they need, and the harder the lungs and heart must work to provide it. This makes your lungs and heart stronger, so they don't have to work as hard when you're not exercising.

Exercising in Space

Astronauts must do special exercises to keep their muscles healthy in the weightless conditions of space. This astronaut, for instance, is working out on an exercise bike. Without gravity to make bones and muscles heavy, the muscles don't have to work very hard. When muscles don't work hard enough, they lose muscle tone and become flabby. Living in space for long periods can seriously weaken muscles.

Speed skaters develop strong gluteus maximus muscles in their buttocks, as well as strong leg muscles.

What Exercise Can Do for You

Certain activities develop particular muscles in your body. Weightlifting and rowing, for example, develop the pectoral muscles in your chest. Playing tennis produces strong arm muscles. Ballet dancing particularly works the muscles in your legs and feet. Singing opera develops your **diaphragm.** Here are some sports and the muscle groups they develop.

From balancing on rings and bars, gymnasts build the deltoid muscles in their shoulders. They also develop strong muscles in their arms and chest.

Swimmers power their strokes with oxygen from their heart and lungs, developing cardiovascular fitness.

Tai Chi

You don't have to move fast to stay fit. These senior citizens in China are practicing the slow-motion art of tai chi, which began thousands of years ago as an exercise for warriors. By stretching and holding parts of your body in a certain sequence, you can use tai chi to build muscle tone and balance. It also improves blood circulation.

Try it!

Improve Your Jump

Do you want to jump higher? You don't need super shoes—just the right technique. The trick is to squat down before you jump. But don't stay down long. Squat and then jump right away. When you squat, your leg muscles stretch out. As you jump up, the muscles contract and shoot you off the ground. It's like shooting a rubber band across the room. If you don't stretch the rubber band, it won't go far. A jump needs that same elastic energy.

Why Do Muscles Cramp?

When you ask your muscles to do work they aren't used to, they will sometimes give you a cramp. A cramp happens when a muscle suddenly contracts and won't let go. The muscle feels tense, like a knot. When you run hard for a long distance, you may get a cramp in your diaphragm, the muscle at the bottom of your rib cage that helps you breathe. These drawings show a relaxed diaphragm muscle *(above, right)* and a cramped one *(right).* Gradually working up to running long distances can help prevent cramps from occurring.

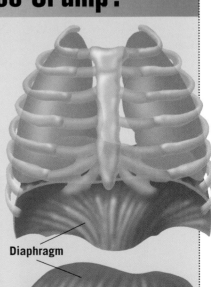

Diaphragm

A Look Inside the Heart

In the center of your chest is a dark red **organ** the size of a fist and no heavier than a couple of baseballs. Yet this small organ is so strong that one squeeze of its **muscle** could send a jet of water 1.8 m (6 ft.) into the air. It is so durable that it beats—contracting and relaxing its muscle—about three billion times in an average lifetime, starting four to six weeks after **conception** and continuing until death. It never stops to rest.

This amazing organ is, of course, the human heart. It is made of a special muscle—cardiac muscle—not found anywhere else in the body. Its job is to pump blood throughout the body. It also serves as a kind of pit stop, steering blood to the lungs for refueling with oxygen before being sent on its way.

The heart is part of the **circulatory system** *(pages 28-29)*, a vast network of vessels that defies gravity to carry life-giving blood both up and down the body and as far as the very tips of your toes and fingers.

The Parts of the Pump

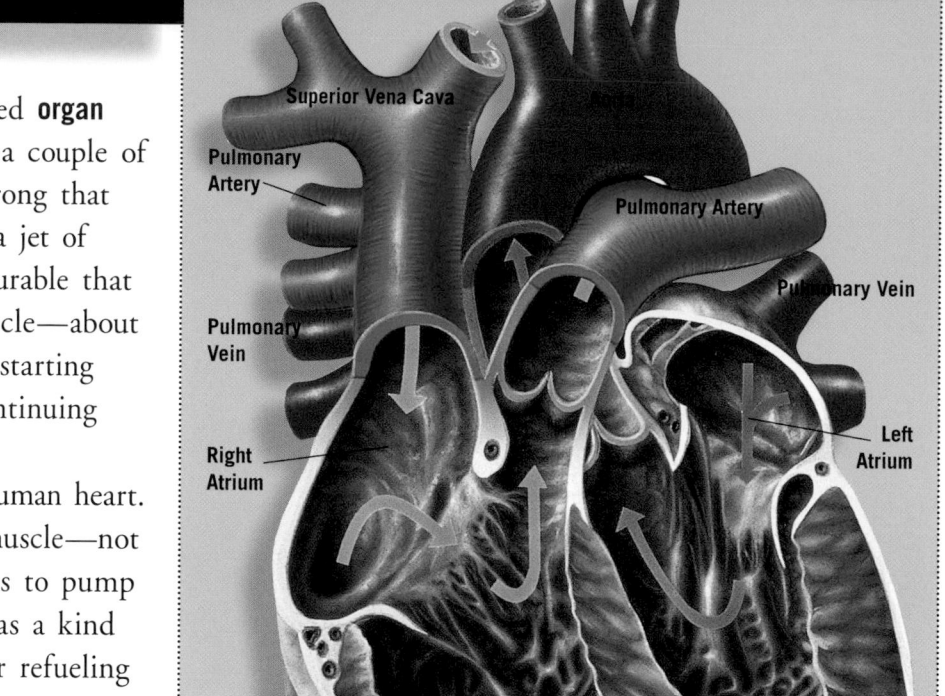

Superior Vena Cava
Aorta
Pulmonary Artery
Pulmonary Artery
Pulmonary Vein
Pulmonary Vein
Right Atrium
Left Atrium
Septum
Inferior Vena Cava
Right Ventricle
Left Ventricle

The heart is a four-part pump. The left and right sides, separated by a wall of muscle (the septum), are divided into two hollow chambers. The upper one is the **atrium,** the lower one the **ventricle.** The heart's right side collects blood returning from the body *(blue arrows)* through the superior and inferior vena cavae. It pumps that blood to the lungs through the pulmonary **arteries.** The left side of the heart receives oxygen-rich blood *(red arrows)* from the lungs through the pulmonary veins and pumps it out to the body through the **aorta.**

The Heart As Art

The first accurate diagrams of the heart appeared about 500 years ago, drawn by the famous Italian artist, scientist, and inventor Leonardo da Vinci (1452-1519). Back then, there were no cameras that could go inside the body and take pictures of a living heart like the one on the opposite page. To learn about the heart, Leonardo and others of his time dissected, or cut open, dead bodies.

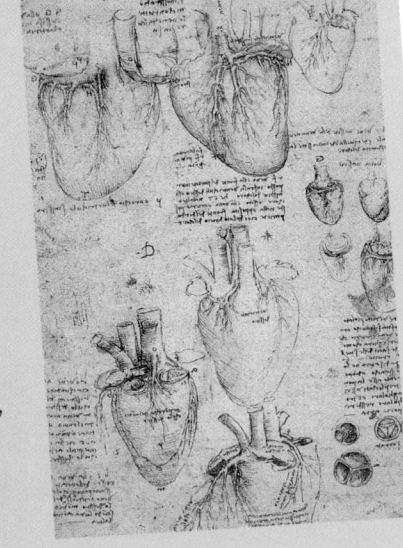

Symbol of Love

The heart symbol has been around for thousands of years. Although its shape is not that of a real heart, it represents to cultures from Swedish to African both the heart and the feeling of love.

Heartstrings

A deep feeling is sometimes said to "tug at the heart-strings." This fanciful phrase refers to **tendons** in the heart that look like tiny cords or strings. In the photo at left you see these tendons inside the right ventricle. They are attached to the tricuspid valve, which prevents blood from flowing backward from the ventricle into the atrium.

Tricuspid Valve

A Steady Beat

A bundle of nerve tissue in the wall of your heart, called the sinoatrial node, or pacemaker, controls the *lub-dub, lub-dub* of your heartbeat. About every second, the pacemaker sends an electrical impulse, or spark, through the heart muscle. This causes both atria to contract and force blood into the ventricles. Then the atrioventricular node shoots out another electrical impulse that contracts both ventricles to shoot blood to the lungs and out into the body.

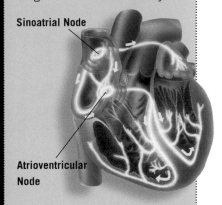

Sinoatrial Node

Atrioventricular Node

Pacemakers

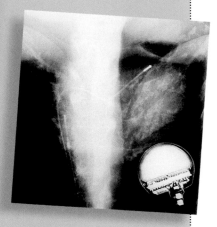

A mechanical pacemaker can often replace a faulty natural one. Worn outside the body, or implanted like the one above, the device is connected to the heart by a wire.

Let's Compare

Hearts come in all shapes and sizes. A grasshopper's heart is a tube. Reptiles and amphibians have two-part hearts that allow oxygen-rich and oxygen-poor blood to mix. Birds and mammals—including humans—have four-chambered hearts that separate the blood going to the lungs from the blood heading for the body.

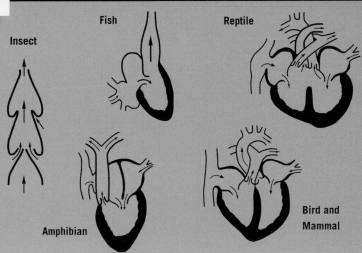

Insect

Fish

Reptile

Amphibian

Bird and Mammal

Pumping Blood

The Circulatory System

Heart

Vein

Artery

In the circulatory system, arteries *(red)* carry oxygenated blood away from the heart. They branch into smaller and smaller vessels until they reach the capillaries. These tiny tubes have thin walls that allow gases and **nutrients** to pass into tissues. Veins *(blue)* then carry oxygen-poor blood back to the heart.

Y our body is caught in a web—a web of blood vessels that weave their way through every inch of your **tissues**. There are so many **veins, arteries,** and **capillaries** in your body, that laid end to end they would stretch more than 96,500 km (60,000 mi.). That's two times around the equator! This huge web of blood vessels, with the heart and blood, makes up the **circulatory system.** Every minute, the heart pumps the body's entire supply of blood—about 5 l (5 qt.) for an adult—through the body. In one day, it pumps more than 7,600 l (2,000 gal.). When blood leaves the heart through the **aorta,** the largest blood vessel, it rockets along at a rate of 40 cm (15 in.) per second. By the time it reaches the tiny blood vessels in the fingers and toes, it slows to 0.05 cm (0.02 in.) per second.

What is Blood Pressure?

T he force of blood pushing against the walls of an artery is the **blood pressure.** Just as water constantly gushing through a hose can wear it out, blood passing through arteries at too high a pressure can damage them. That's why it is important to have your blood pressure checked. When a doctor or nurse measures your blood pressure, he or she says two numbers, such as "120 over 80" (normal for an adult). The first number is the systolic pressure, the pressure of the contracted heart squeezing blood out of the heart. The second is the diastolic pressure, the pressure when the heart relaxes and fills with blood.

A sphygmomanometer is an instrument that measures blood pressure in millimeters of mercury. A stethoscope is used to hear the heart's pulse.

The Heart at Work

1

Blood that has circulated through the body, losing its oxygen and collecting carbon dioxide, enters the right **atrium** of the heart through the vena cava. When the atrium contracts, it pushes the blood through the tricuspid valve and into the right **ventricle.**

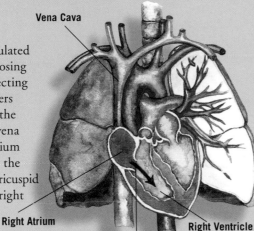

Vena Cava

Right Atrium

Tricuspid Valve

Right Ventricle

2

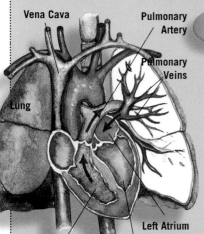

Vena Cava

Pulmonary Artery

Pulmonary Veins

Lung

Right Ventricle

Left Atrium

Left Ventricle

When the right ventricle contracts, it pushes blood through the pulmonary artery into the lungs. There, tiny blood vessels called capillaries absorb carbon dioxide from the blood and replace it with oxygen. Then the freshly oxygenated blood flows from the lungs into the left atrium through the pulmonary veins.

3

From the left atrium, the oxygen-rich blood is pumped through the mitral valve and into the left ventricle. Here the heart **muscle** contracts the strongest. Blood shoots out from the left ventricle through the aorta and out to the body. After its journey through the body, it will return to the heart and the cycle will start all over again.

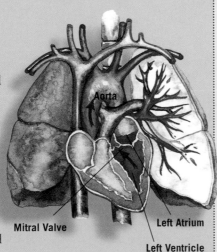

Aorta

Mitral Valve

Left Atrium

Left Ventricle

Blood Vess[els]

Blood vessels come in all different sizes, depending on their function. The largest ones, arteries, can be as thick as a thumb. They branch off into smaller vessels, called **arterioles,** which connect to capillaries, the tiniest and most abundant blood vessels *(photograph).*

From the capillaries, blood passes into slightly larger vessels called **venules.** These join with veins, which carry blood back to the heart. Veins, which have lower blood pres[sure] than arteries, are the only vessels with valves that prevent blood from flowing backward.

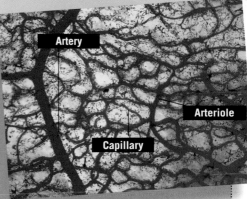

Artery

Arteriole

Capillary

People — William Harvey

Before the 17th century, people weren't sure how the circulatory system worked. Some Greeks thought blood vessels carried air. Others thought blood ebbed and flowed like an ocean tide.

The English physician William Harvey *(below, right)* finally set things straight. He proved that blood flows in one direction in a continuous circle. In experiments, Harvey pressed on veins *(below, left)* to show that blood in veins always flows toward the heart and will not run backward. His discoveries became the basis for modern research on the heart.

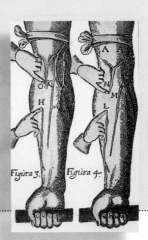

Figura 3. Figura 4.

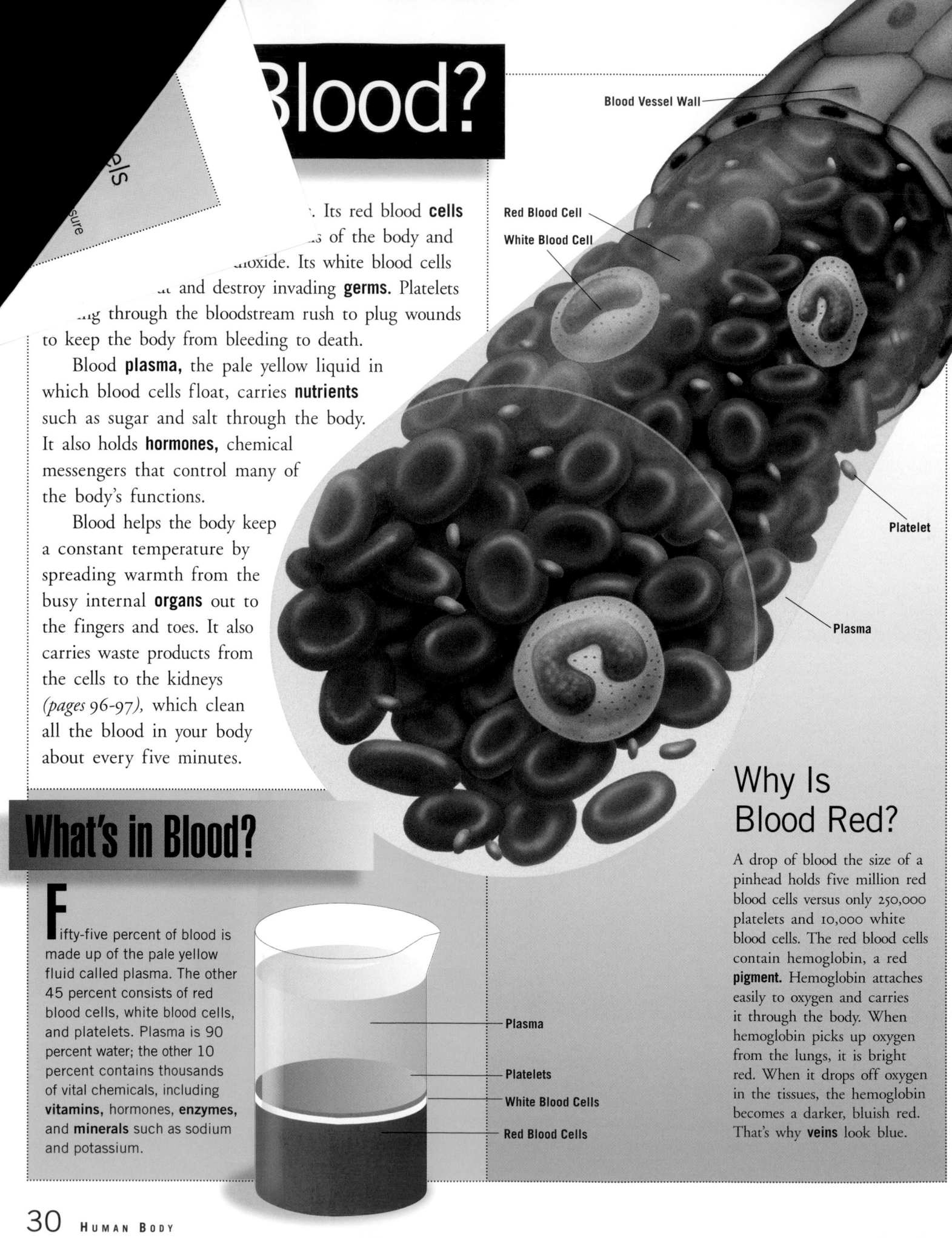

Blood?

. Its red blood **cells** ... s of the body and ... ...oxide. Its white blood cells ... .t and destroy invading **germs.** Platelets ... ...ng through the bloodstream rush to plug wounds to keep the body from bleeding to death.

Blood **plasma,** the pale yellow liquid in which blood cells float, carries **nutrients** such as sugar and salt through the body. It also holds **hormones,** chemical messengers that control many of the body's functions.

Blood helps the body keep a constant temperature by spreading warmth from the busy internal **organs** out to the fingers and toes. It also carries waste products from the cells to the kidneys *(pages 96-97),* which clean all the blood in your body about every five minutes.

Blood Vessel Wall

Red Blood Cell

White Blood Cell

Platelet

Plasma

What's in Blood?

Fifty-five percent of blood is made up of the pale yellow fluid called plasma. The other 45 percent consists of red blood cells, white blood cells, and platelets. Plasma is 90 percent water; the other 10 percent contains thousands of vital chemicals, including **vitamins,** hormones, **enzymes,** and **minerals** such as sodium and potassium.

Plasma

Platelets

White Blood Cells

Red Blood Cells

Why Is Blood Red?

A drop of blood the size of a pinhead holds five million red blood cells versus only 250,000 platelets and 10,000 white blood cells. The red blood cells contain hemoglobin, a red **pigment.** Hemoglobin attaches easily to oxygen and carries it through the body. When hemoglobin picks up oxygen from the lungs, it is bright red. When it drops off oxygen in the tissues, the hemoglobin becomes a darker, bluish red. That's why **veins** look blue.

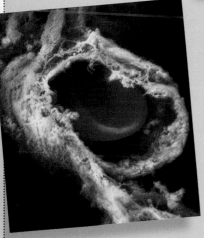

Capillaries are the tiniest blood vessels in our body. They are so narrow that red blood cells—some of the smallest cells in our body—must pass through them one cell at a time *(left)*. Even though they are tiny, **capillaries** have a big job to do.

Why Do We Get Flushed?

When we exercise, we burn energy, which creates heat. To keep our body from over-heating, blood carries this heat to the skin's surface, where it radiates into the air. The skin of the face has lots of capillaries close to the surface. When you get hot, these capillaries swell with blood, making them bigger and your face redder. That's why your face looks flushed.

Where Is Blood Made?

Blood cells have short lives. Red blood cells live for 100 to 125 days, platelets about 10 days, and white blood cells only six to nine days. So where do new blood cells come from?

Most blood cells are made in the marrow, or spongy **tissue**, inside the bones. In a newborn baby, all the bones are involved in making blood cells. In adults, only the long bones, such as the femur *(below, left)*, and the flat bones, like the **sternum**, make red blood cells, platelets, and some kinds of white blood cells. Lymph nodes *(page 98)*, special clumps of tissue found throughout the body, make other white blood cells.

Bone marrow (below), a spongy tissue inside bones, churns out most blood cells.

Bone Marrow

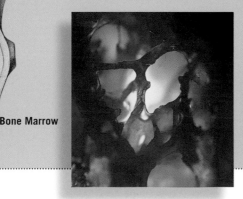

Strange But TRUE! Blue Blood

Not all blood is red. Lobsters, for example, have blue blood. Instead of reddish, iron-based hemoglobin in their blood, they have hemocyanin, a bluish pigment that contains copper.

What Can Blood Tell Us?

Blood travels through the whole body, so even a tiny sample of it can yield valuable clues about a person's health. A group of tests called a complete blood count checks the number of red cells, the hemoglobin content, and the number of white blood cells in a blood sample. A doctor will then compare these numbers with a standard for healthy blood. Too many white blood cells may indicate infection; too little hemoglobin might mean the patient is anemic, or low in iron. Hundreds of other blood tests can pick up more specific problems.

Blood Close up

An amazing army of natural defenses spring into action when you cut yourself. To seal a cut and help it heal, a web of fibrin strands *(right)* forms in the **plasma.** The strands trap red blood **cells** and platelets to create a clot. On the surface of the skin, this clotting forms a scab. The process is explained in the illustrations below.

Repairing a Wound

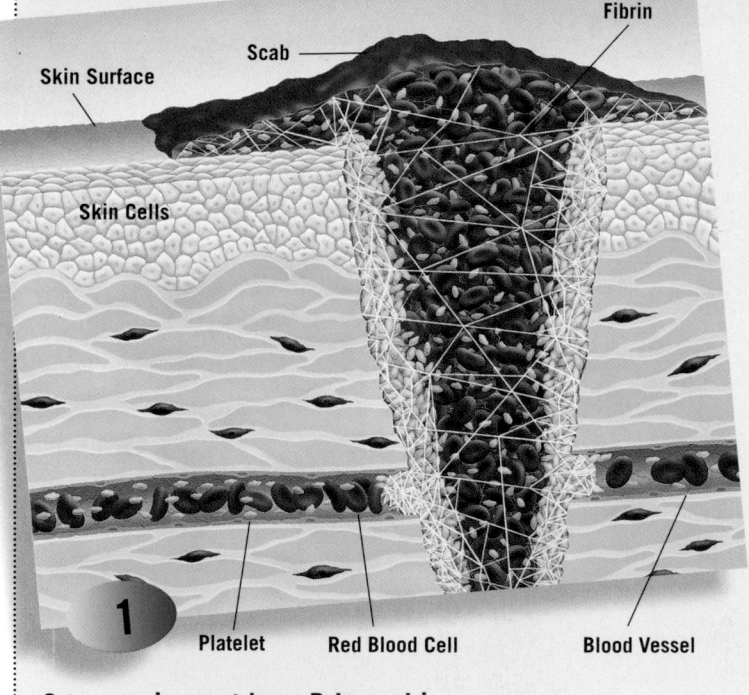

Skin Surface
Scab
Fibrin
Skin Cells

1

Platelet
Red Blood Cell
Blood Vessel

Stopping the Bleeding

When a cut bleeds, platelets cluster along the edges of the wound, plugging the opening. They also release chemicals that help make a **protein** called fibrin. Fibrin forms long threads in a crisscross pattern across the wound. The threads trap red blood cells and platelets, strengthening the blood clot.

2

Forming a Scab

The blood clot dries into a hard scab on the skin's surface. Beneath it, skin cells grow and form a bridge across the wound. The damaged blood vessel starts to rebuild itself. **Macrophages,** a kind of white blood cell, devour dead and injured cells as well as **germs.**

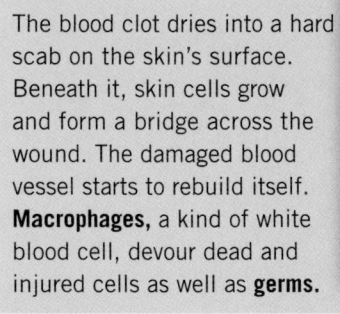

Macrophage

3

Healing the Skin

The skin gradually returns to its normal thickness, causing the scab to loosen and fall off. The blood vessel is completely repaired, although it doesn't necessarily follow the same path as before. Here it forms an arch where the cut was.

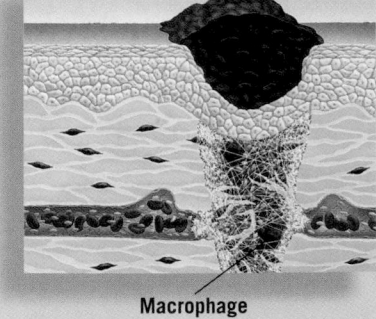

Then & NOW!

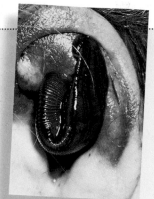

For much of history, people have thought that sickness was caused by bad blood—and that the cure was bloodletting, or draining blood. Often the loss of blood from an opened **vein** or a blood-

sucking leech *(left)* actually weakened or even killed the patient. The practice was finally abandoned about the beginning of the 20th century. Today bloodletting is back. Doctors are using leeches after operations in which they reconnect detached fingers and ears *(above)*. Leech **saliva** prevents blood from clotting, so blood flows freely through the reattached body part, allowing the wound to heal.

Blood Transfusions

The first successful blood transfusion reportedly happened in the 17th century. French physician Jean-Baptiste Denis drained lamb's blood into the arm of a feverish boy. Miraculously, the boy survived. Denis's second patient, shown in this 1667 print *(below)*, was not so lucky. He died while celebrating in a tavern shortly after the transfusion. What was then a risky procedure was soon outlawed. Only after the discovery of blood groups and blood typing in the 20th century were transfusions considered safe.

Blood Types

There are four types of blood—type O, type A, type B, and type AB. A simple blood test tells what type you have. Some blood types match, but others don't. If a person receives the wrong type, the blood will thicken and the person will eventually die.

Type A blood can receive type A and type O blood.

Type B blood is compatible with itself and with type O blood.

Type AB blood can receive any blood type.

Type O blood, the "universal donor," can donate to all types but can receive only its own type.

People · Charles R. Drew

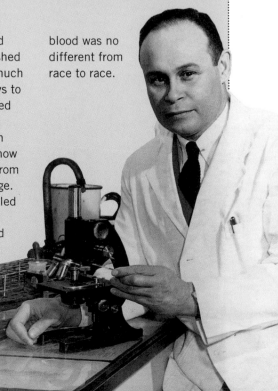

By the late 1930s, blood transfusion was an established procedure, but there was much to learn about the best ways to preserve and handle donated blood. Charles R. Drew, an African-American physician and professor, discovered how to separate blood plasma from blood cells for better storage. During World War II, Drew led a national blood collection program for the U.S. armed forces. He resigned when "black" blood was segregated from "white" blood, in spite of scientific evidence that proved human blood was no different from race to race.

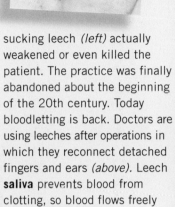
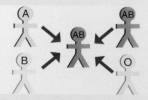

Keeping the Beat

With every heartbeat, blood surges through the **circulatory system.** You can feel this regular surge—or **pulse**—at certain places on your body where **arteries** are close to the skin. The pulse is a measure of your heart rate, or how fast your heart is beating.

The smaller the body, the faster the heart beats. Babies, for example, have a pulse of about 120 beats per minute (bpm). A child's heart beats between 80 and 100 bpm. The normal heart rate for an adult is 72 bpm. That equals 4,320 beats per hour and 103,680 beats each day. At that rate, in a year the heart beats 37,843,200 times!

Animals have different heart rates, too. The heart of a shrew, one of the smallest mammals, races at almost 1,000 bpm. An elephant's heart beats with a slow thunder, only 20 to 30 times per minute.

Pulse Points

The body has many pulse points (right). You can check your pulse by pressing lightly on any of these spots. The easiest place to take your pulse is probably in the wrist. Place two or three fingers of one hand on the wrist, below the thumb, and press gently (below). Count the pulses for 20 seconds and multiply that number by 3. That's how many times your heart is beating per minute.

Diagnosis by Pulses

Doctors in ancient China believed that they could find the source of an illness by making a detailed study of the pulse. By the sixth or fifth century BC, they had associated specific characteristics of the pulse with particular internal **organs.** A pulse chart, like this one from 1693, helped them make a diagnosis. The chart describes many possible readings of the pulse and identifies the medical problem connected with each reading.

Today, doctors use changes in the pulse's rate, rhythm, and strength to help diagnose health problems. Even in ancient times, people understood that the pulse was a valuable clue to sickness.

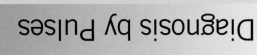

Changing the Beat

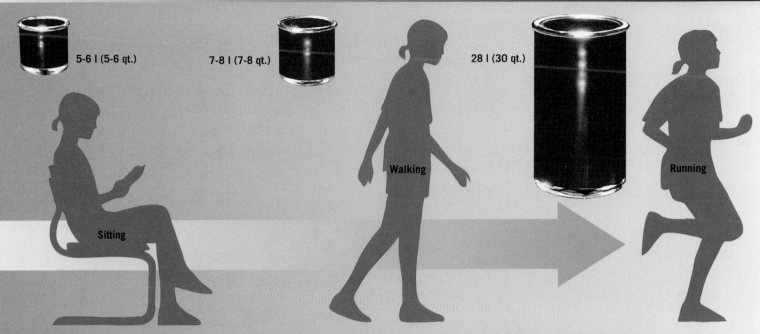

5-6 l (5-6 qt.)

7-8 l (7-8 qt.)

28 l (30 qt.)

Sitting

Walking

Running

When the body works hard, the heart works hard to give it more oxygen. At rest, an adult's heart beats 60 to 80 times and pumps 5 to 6 l (5 to 6 qt.) of blood each minute. Walking increases the heart rate to 100 to 120 beats per minute and raises blood flow to 7 to 8 l (7 to 8 qt.) per minute. Strenuous exercise may double the number of beats per minute to 200, making the heart pump an amazing 28 l (30 qt.) per minute.

Hardy Heart

The heart **muscle** is remarkably strong. It exerts enough force in one hour to lift 1,350 kg (3,000 lb.)—about the weight of a small car—0.30 m (1 ft.) off the ground.

The Heart Never Lies

When a person tells a lie, the body—especially the heart—responds. **Blood pressure,** perspiration, and heart rate may rise. Sometimes breathing speeds up.

A lie detector, or polygraph, is a set of instruments that measure and record these responses. An operator winds a strap around the person's chest and wraps a blood-pressure cuff around the arm. He or she asks the person a series of questions, while the instruments record the body's reactions to them. The data are then fed into a computer and displayed on the screen (below).

Drugs and emotions such as anger, pain, and fear can cause similar bodily changes, so a lie detector test is not always right.

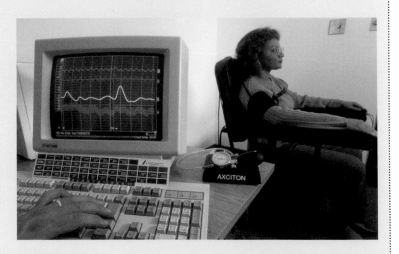

What Is a Heart Attack?

It begins with pain—severe chest pain and possibly more pain in the jaw, neck, arm, or back. The pain may last several minutes or several hours. Some people who have experienced it say it feels as though a giant fist is squeezing the heart. Others don't feel any pain, but they do break out in a cold sweat, feel faint, and perhaps have indigestion. It's time to get to a hospital—fast. These are symptoms of a heart attack.

Heart attacks usually happen when the heart **muscle** is starved for blood. Two large, branching coronary **arteries** deliver oxygenated blood and **nutrients** to the **tissues** of the heart. Over time, these arteries may get damaged or clogged. When this happens, the heart may stop beating, or beat only weakly. This deprives the body of blood. It is a medical emergency.

Clogged Coronary Artery

Dead Heart Muscle

What's an EKG?

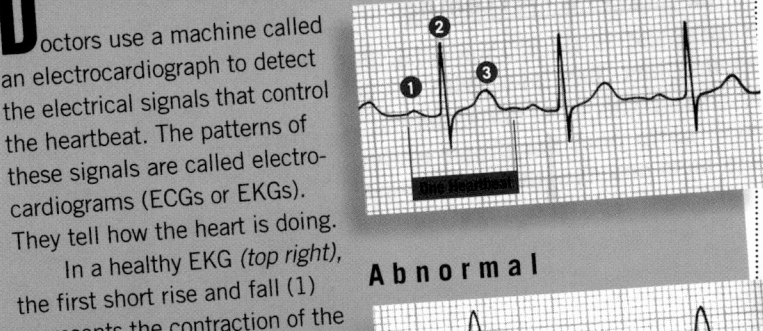

Doctors use a machine called an electrocardiograph to detect the electrical signals that control the heartbeat. The patterns of these signals are called electrocardiograms (ECGs or EKGs). They tell how the heart is doing. In a healthy EKG (top right), the first short rise and fall (1) represents the contraction of the **atria.** The tall peak and drop (2) is the stronger contraction of the **ventricles,** called systole. The short peak (3) that follows shows that the ventricles are relaxed and the heart is filling with blood. This is diastole. Irregularities in this pattern (bottom right) may mean heart problems.

Normal

Abnormal

What Causes an Attack?

Most heart attacks begin with a clog in one of the coronary arteries, the vessels that supply blood to the heart muscle. No blood, and therefore no oxygen, can get through to the muscle, so a section of the heart muscle dies. The weakened heart cannot pump enough blood through the body. If too much heart muscle dies, a person cannot survive.

Arteries are clogged when a fatty substance called **cholesterol** builds up and hardens on an artery wall.

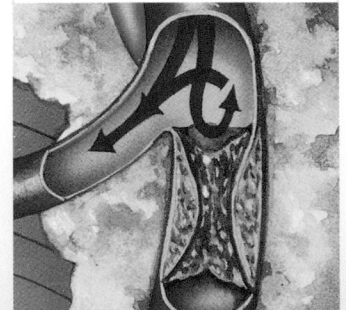

Sometimes a blood clot forms over the cholesterol and blocks the artery (above). Then blood can't get through, and the heart muscle is deprived of oxygen.

Creating a Clog

Fat clings to the wall of an artery in the photo below. This fat can block the artery, or float to another part of the body and choke a different blood vessel, damaging other tissues.

Rebooting the Heart

Sometimes doctors can control an irregular heartbeat, or restart a heart if it stops. For example, a medicine called digitalis increases the heart muscle's strength, improving blood flow. And the heart can often be jump-started by a defibrillator. This instrument sends an electrical current through the chest, causing all the chest muscles, including the heart, to contract.

Digitalis is made using the leaves and seeds of the foxglove plant (above).

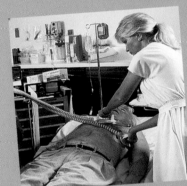

A defibrillator (left) sends an electric shock to the chest muscles, triggering a normal heartbeat.

Strange But TRUE!

Heart-Stopping Rescues

Normally, if the heart stops beating, death follows within about 15 minutes. But 11-year-old Alvaro Garza Jr., shown here being rescued from an icy river in North Dakota, survived after being clinically dead for 45 minutes. Even more incredible, 51-year-old Norwegian fisherman Jan Egil Refsdahl was revived after his heart had stopped beating for four hours! What saved these lucky people was the cold. Both fell into icy waters that lowered their body temperature a lot. When this happens, the body automatically shuts down all functions except those most necessary for life. The heart slows way down. Blood flow to the skin is reduced as the body hoards warm blood for vital **organs,** such as the brain. In these rare cases, the body can survive with little or no heartbeat.

Mending a Broken Heart

The human body is a complex machine, and as with any machine, sometimes its parts break down. For a long time, people thought the heart was too complicated and delicate to fix. But today there are lots of ways in which cardiologists—doctors who specialize in the heart—can fix a broken heart.

If the heart **muscle** isn't getting enough blood because a coronary **artery** is blocked *(page 36)*, sometimes medicine can dissolve the blockage. At other times, doctors can widen the vessel and clean it out. If necessary, they can even bypass the damaged part of the vessel and create a new path for the blood.

When the heart can't be repaired, it must be replaced. A new heart can be transplanted from a person who has just died. However, there aren't enough donors for all the people who need a new heart. Therefore, researchers are looking at ways to transplant animal hearts or make mechanical ones.

Detour for the Blood

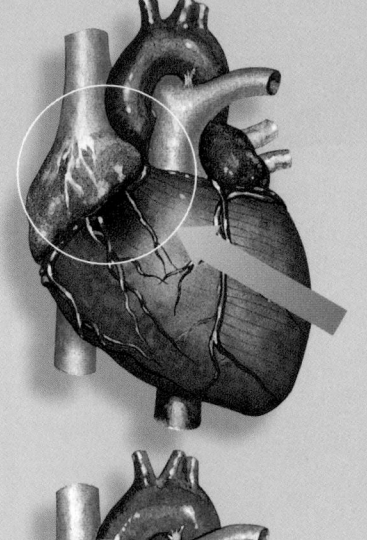
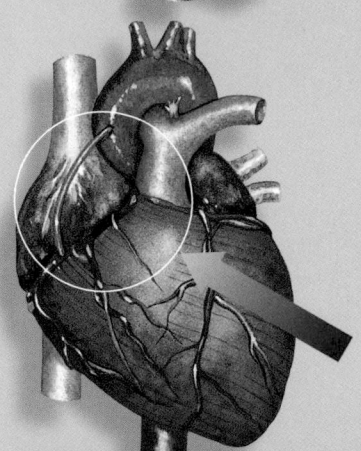

When an obstacle blocks traffic, police set up a detour so cars can zoom around the trouble spot and keep going. Bypass surgery can do the same thing for the heart. If a fatty deposit is slowing or even stopping the flow of blood to the heart muscle *(top left)*, doctors can build a detour for the blood. Removing a section of an artery or **vein** from another part of the body, they attach one end of it to the **aorta.** Then they attach the other end to the blocked artery below the clog. Using this new route, blood returns to the deprived heart muscle *(bottom left)*.

The Balloon Solution

When fatty deposits, called plaque, block an artery in the heart, sometimes the vessel can be cleared by a procedure called balloon angioplasty. A doctor slides

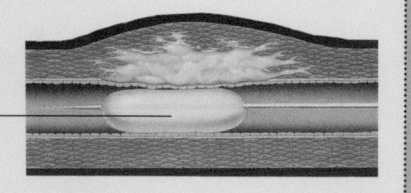

The deflated balloon reaches the plaque.

The inflated balloon opens the artery, pressing the plaque into the vessel wall.

a very thin tube with a tiny, deflated balloon on the end through the blood vessel. When it reaches the blocked area, the doctor inflates the balloon. This squeezes the plaque against the sides of the vessel, stretching the artery so there is room for the blood to flow by.

People Daniel Hale Williams

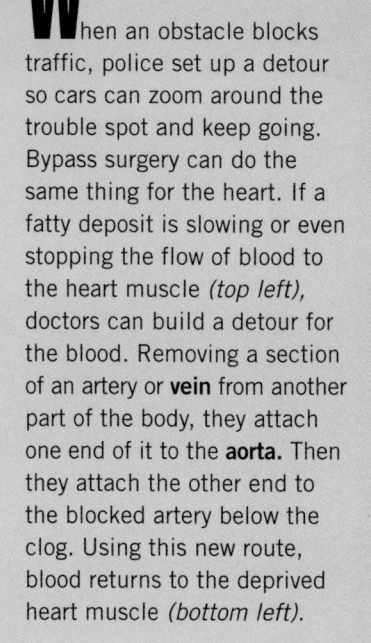

In 1893, a man was rushed to Provident Hospital in Chicago with a stab wound to the heart. In those days, a heart wound was usually fatal. But Dr. Daniel Hale Williams, the African-American doctor who founded the hospital, decided to operate. Fifty-one days later, the patient left the hospital whole and healthy, the survivor of the first successful heart surgery.

Williams was a pioneer in many ways. In 1913, he became a surgeon at a previously all-white hospital in Chicago. He was then named a charter member of the American College of Surgeons, the only black among its 100 founders.

Mechanical Parts

Heart valves keep blood from flowing backward through the heart. Sometimes their normally flexible flaps get stiff or weak, no longer forming a tight seal. Then the valves may be replaced by artificial ones. You can see one in the x-ray of a chest at right (seen from the side). Artificial valves like the ones below are made of metal and plastic. Valves may also be made from organic material such as **tissue** from a calf's heart.

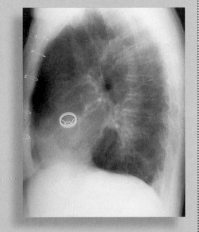

The Gift of Life

People who agree to donate their heart and other **organs** when they die save many lives each year. One such life is that of Andrea Mongiardo, pictured below showing off his new strength. Andrea, an Italian boy, was born with only one **ventricle.** By 1994, he was 15 years old and his body was failing. When an American boy named Nicholas Green died suddenly while vacationing in Italy, his parents decided to donate his organs so that others might be able to live. Andrea was the lucky recipient of Nicholas's heart.

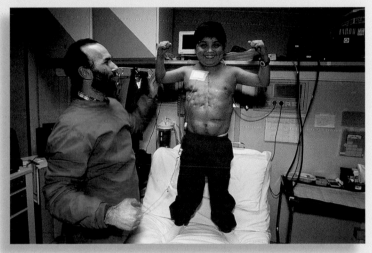

Famous 1 FIRSTS

Human Heart Transplant

In December 1967, South African heart surgeon Christiaan Barnard had a patient with heavily clogged coronary arteries. Unable to repair the

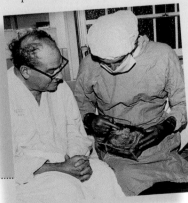

heart, Dr. Barnard decided to try something new. He removed the patient's sick heart and replaced it with a healthy one, donated by the family of a woman who had recently been killed in an accident. Dr. Barnard had performed the first successful human heart transplant. Two months later he transplanted another heart. At left, Dr. Barnard shows his second transplant patient the diseased heart he just removed from his body.

Strange But TRUE!

Pigs to the Rescue

There are far more people in need of new hearts or other organs than there are human organs available. This is a problem pigs may be able to help solve.

Pigs have organs that are very similar to those of humans. Even so, a human body will probably reject a pig's organ unless scientists can find a way to trick the body into accepting it. Some researchers now hope to do this by injecting pigs' eggs with human **genetic** material. In this way they may breed animals with organs that are more like a person's.

Heart History

A ncient people knew that the heart was important—but they didn't always know why. Ancient Egyptians believed the heart was the center of emotions and the intellect. They knew that blood vessels started in the heart and linked it to the rest of the body. But they also believed that these vessels carried fluids like tears and urine along with blood.

The Greek philosophers Plato and Aristotle thought the heart was a "vital flame" in which food was burned. This flame created life and warmth. Breathing cooled the flame and stopped it from burning too high and consuming the body.

Amid all the misunderstandings, sometimes people explained it right. A Chinese medical book from about 3,000 years ago correctly reported that the heart regulates all the blood of the body and that blood flows continuously in a circle. Westerners did not reach the same conclusion until the 17th century.

A Necessary Sacrifice

The Aztecs believed that they needed to give hearts and blood to the gods to keep the universe in order. Particularly demanding was the war god Huitzilopochtli, who required a daily sacrifice. The Aztecs thought he was the sun. They feared he would not rise in the morning if he didn't get a sacrifice. In the illustration below, a priest cuts out the heart of a sacrificed captive. The heart rises heavenward. In reality it would have been placed in a special container.

Heart and Soul

The ancient Egyptians believed that death was the gateway to a glorious afterlife—for those who truly deserved it. The heart would decide who was worthy. According to Egyptian lore, the jackal-headed god Anubis would place the dead person's heart on a scale and weigh it against a feather from the headdress of Maat, goddess of order. If the scale balanced perfectly, the dead person was welcomed into the afterlife. If it did not, the monster of the dead, Ammit *(squatting next to scale)*, gobbled it up.

Hippocrates

Hippocrates, often called the father of medicine, was a Greek teacher and traveling doctor who lived in the fifth century BC. His name is given to a group of medical writings called the *Hippocratic Collection,* although he probably wrote only some of them.

This collection correctly describes the heart as having **valves, atria** and **ventricles** that contract at different times, and great blood vessels extending from it. But the ancient Greeks didn't under-stand the differences between **arteries** and **veins.** They also thought that air traveled through the blood vessels along with blood.

Galen's Gospel

In the second century, the Greek physician Galen published anatomy and physiology texts that would be followed for the next 1,400 years.

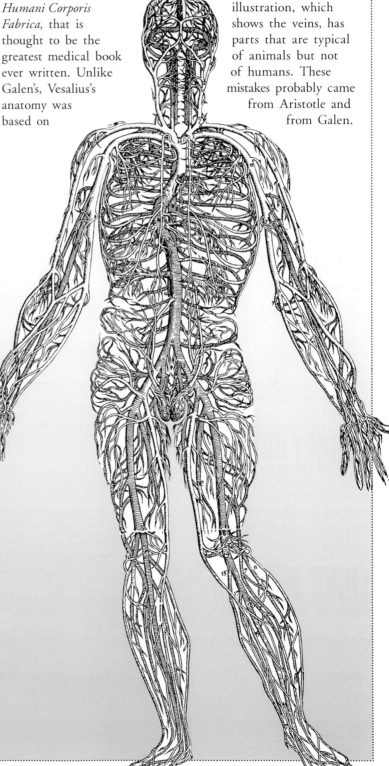

He discovered some important facts—for instance, that blood circulates through the heart and lungs—but he made some mistakes as well.

These were perpetuated through time, as shown by this drawing taken from a 13th-century English medical manuscript. It pictures the **esophagus** as leading directly to the heart, not the stomach.

Vesalius

In the 16th century, the Flemish physician Andreas Vesalius published an anatomy text, *De Humani Corporis Fabrica,* that is thought to be the greatest medical book ever written. Unlike Galen's, Vesalius's anatomy was based on studies of human bodies, not on faith and animal dissection. Even so, his findings still had some errors. This illustration, which shows the veins, has parts that are typical of animals but not of humans. These mistakes probably came from Aristotle and from Galen.

The Lungs

The Inside Story

You can live for weeks without food. You can survive for a few days without water. But without air, you will die in a few minutes. This is because the **cells** in your body depend on oxygen, a gas found in the air, to turn food into energy.

When you breathe in, air enters your **respiratory system,** which begins with your mouth and nose. The air travels down your throat through a tube called the trachea. At the end of the trachea, the airway splits in two. These are the bronchi. One bronchus carries air to the left lung, and the other carries it to the right. Inside the lungs, the bronchi branch out into many smaller air tubes called bronchioles.

Tiny blood vessels in the lungs pick up oxygen from the air and release carbon dioxide produced by your body. The carbon dioxide—along with water vapor and heat—leaves your body when you breathe out.

The delicate, pink, spongy lungs are protected in the chest by a cage of bones—the ribs, the **sternum** (breastbone), and the vertebral column (backbone). The lungs rest on a **muscle** called the **diaphragm,** which moves up and down to help the lungs pull air into and push air out of the body.

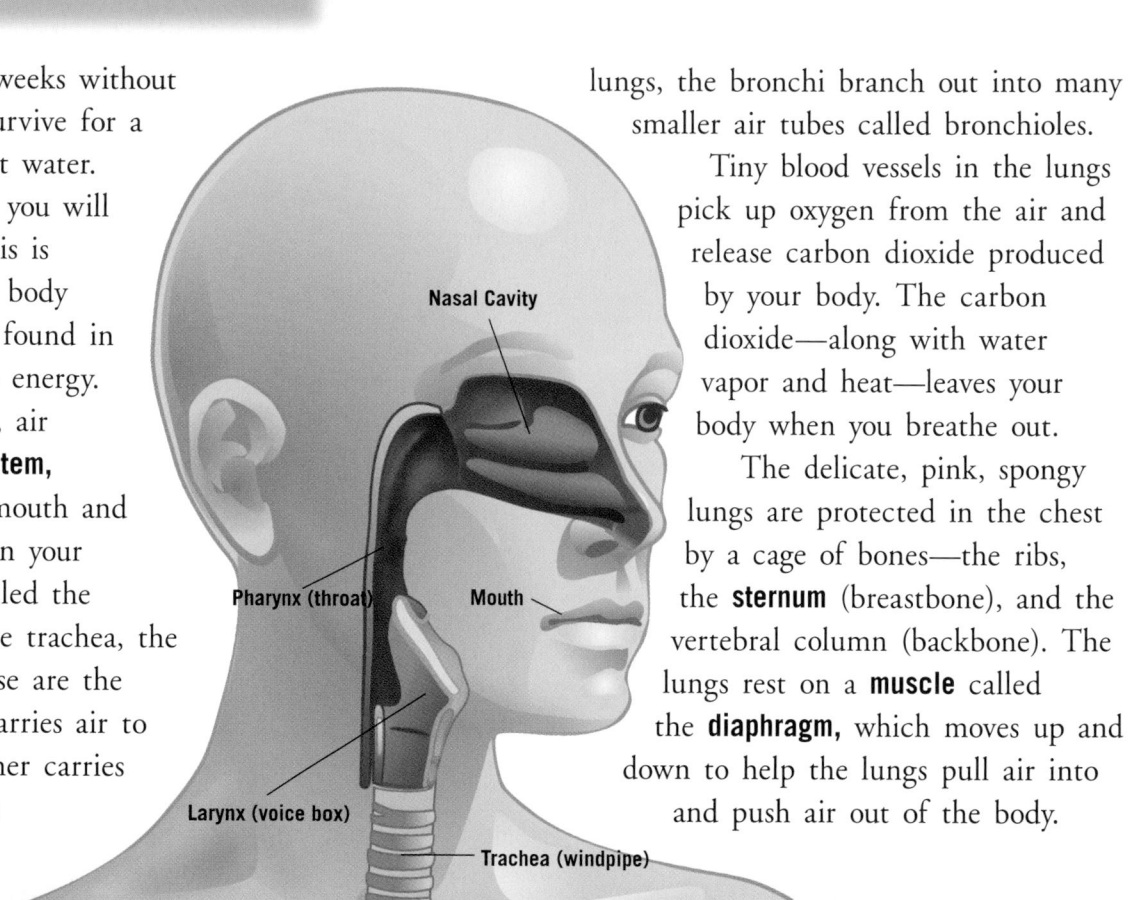

Nasal Cavity

Pharynx (throat)

Mouth

Larynx (voice box)

Trachea (windpipe)

Bronchioles

Bronchus

Right Lung

Left Lung

Diaphragm

Making the Exchange

Capillaries

Alveoli

You have a tree inside your lungs— a beautiful, branching structure of air tubes known as the bronchial tree *(left)*. The main branches, or bronchi, split into smaller and smaller air tubes. Eventually the tubes are thinner than a human hair. These are called bronchioles.

At each bronchiole's end are clusters of air sacs, called alveoli, that look like bunches of grapes. Wrapped around each sac is a net of **capillaries**. Oxygen passes from the bronchioles into the alveoli, and from there into the capillaries. The bloodstream then carries the oxygen to cells throughout the body. In a reverse process, the alveoli pick up waste carbon dioxide from blood in the capillaries and expel it when you breathe out.

Alveoli

Lungs have an amazingly compact and efficient design. Most of the space inside them is taken up by tiny alveoli, only about 0.02 cm (0.008 in.) across. About 300 million of these air sacs fill the lungs. Spread out flat, they would cover a surface as large as a tennis court.

The alveoli need to have such a large surface area in order to handle the body's constant need to exchange gases. Each alveolus fills and empties about 24,000 times in a day of normal breathing.

Polluting the Lungs

A newborn baby has healthy pink lungs. As he or she ages, the lungs gradually turn gray or even black. This is because they collect small bits of dust and dirt from the air. People who live in cities often have blacker lungs than people who live in less polluted areas.

Smoking causes some of the worst damage to lungs. Tar and soot from smoke stick to the insides of the air sacs and air passages, as in the picture at right. This sticky dirt irritates the airways and forces the body to work harder to get enough oxygen.

Let's Compare

Lungs and Gills

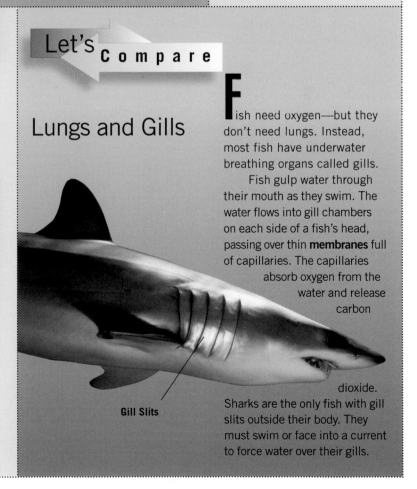

Fish need oxygen—but they don't need lungs. Instead, most fish have underwater breathing organs called gills.

Fish gulp water through their mouth as they swim. The water flows into gill chambers on each side of a fish's head, passing over thin **membranes** full of capillaries. The capillaries absorb oxygen from the water and release carbon dioxide.

Sharks are the only fish with gill slits outside their body. They must swim or face into a current to force water over their gills.

Gill Slits

How Do We **Breathe?**

Taking a Breath

You don't usually have to think about breathing. It happens automatically, thanks to a signal sent from the brain. The breathing control center is located in the medulla oblongata, where the **spinal cord** meets the base of the brain *(pages 50-51).*

Because lungs cannot move by themselves, breathing depends on two kinds of **muscles:** the **diaphragm,** the dome-shaped muscle below the lungs, and the intercostal muscles of the chest wall between the ribs.

When you inhale, these muscles contract. The diaphragm pulls downward and the rib cage expands, stretching the lungs and causing them to suck in air. When you exhale, the intercostal muscles relax, and the chest shrinks. The diaphragm relaxes and rises up. The elastic lungs spring back to their original size, forcing the air out.

Fast FACTS

In a day, the average person breathes about 24,000 times. In a 70-year lifetime, that's some 600 million breaths.

You will inhale about 390,000 cu m (13 million cu ft.) of air during a lifetime. That's enough to fill about 52½ Goodyear blimps.

A man's lungs can hold 6 l (6.4 qt.) of air; a woman's hold about 4.2 l (4.5 qt.).

A 10-year-old at rest inhales and exhales about 20 times per minute. An adult breathes in and out six to 12 times per minute.

You need oxygen to live, but too much of it can kill you. If you breathe in 100 percent oxygen (five times as much as you find in air) for too long, your lungs can fill with fluid and your air sacs can collapse.

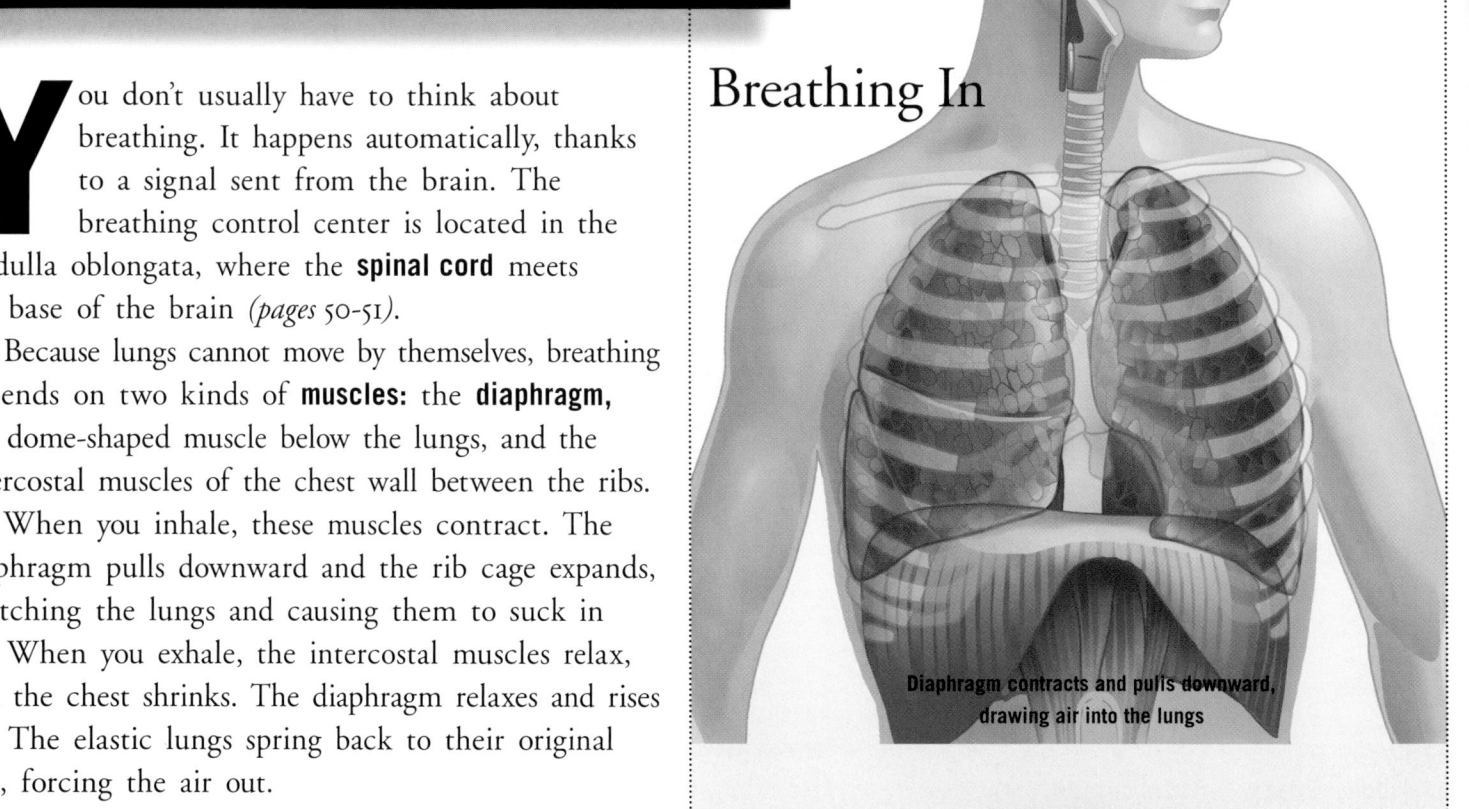

Breathing In

Diaphragm contracts and pulls downward, drawing air into the lungs

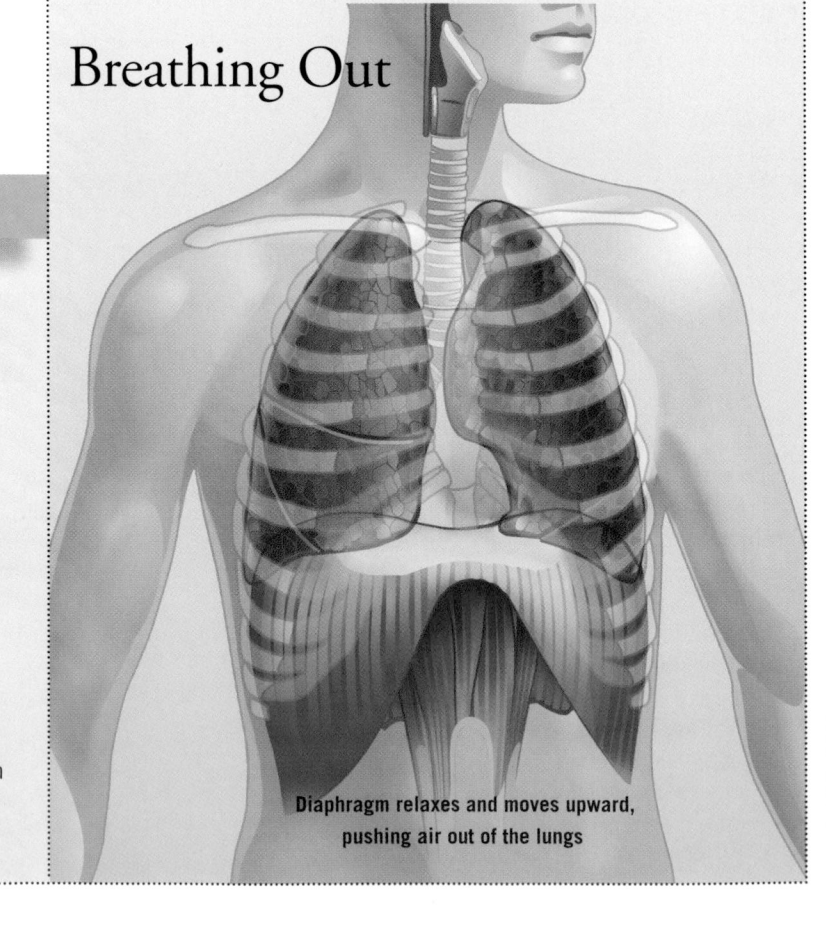

Breathing Out

Diaphragm relaxes and moves upward, pushing air out of the lungs

I Was There!

"The higher you go, the thinner the air. Kind of like whole milk and skim milk. If you're used to drinking whole milk, then skim milk's going to taste awful, almost like water. It takes getting used to. The same for air. The thinner the air, the less oxygen it has. If your body is used to sea-level air, it takes a while to get used to the thin air up high. If you're too active when you first get to higher elevations, you'll get headaches and gradually get pretty sick to your stomach. If you acclimatize well, slowly enough to adjust but not so slowly you are bored to death, you can keep moving up higher and higher. . . . Otherwise, the altitude will get you every time."

In 1996, 16-year-old Mark Pfetzer became the youngest person ever to attempt Mount Everest. In a book about his experience, he describes *(left)* what it feels like to be 8,000 m (26,000 ft.) high, where there is so little oxygen that you have to wear an oxygen tank.

Instead of a larynx, or voice box, birds have a syrinx, or song box. In songbirds, the syrinx is very elaborate. Some have as many as six pairs of muscles to control the specialized membranes and walls of the syrinx. By controlling the right and left sides of the syrinx separately, songbirds can sing two different melodies at the same time.

The Human Voice

When you breathe, air rushes down the trachea and passes through the larynx, or voice box. On either side of the larynx are two cords, called vocal cords, made of a tough, skinlike substance. If you contract the muscles in your neck, the vocal cords close and form a narrow gap. Air traveling through this gap vibrates the cords and makes a sound—the human voice.

When a boy reaches puberty, **hormones** enlarge his larynx, lengthening the cords. His voice drops down in pitch. Girls go through a similar, though less dramatic, change.

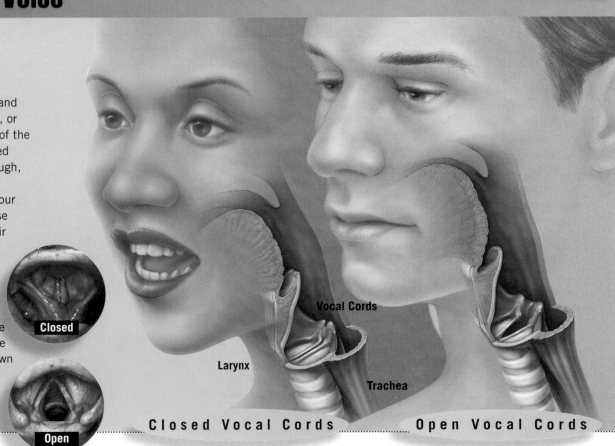

Closed

Open

Vocal Cords

Larynx

Trachea

Closed Vocal Cords

Open Vocal Cords

Why Do We Sneeze?

t starts with a little tickle in your nose. Then it seems nothing can stop you—you've got to sneeze.

Sneezing is such a powerful event that some ancient cultures thought that you lost your soul, along with your breath, when you sneezed. Even worse, the violent expulsion of "soul" made room for demons to rush in.

What's really going on when you sneeze? A sneeze is a **reflex** *(page 49)* triggered when something irritates the lining of the nose. The **muscles** involved with breathing involuntarily contract, suddenly and violently forcing air out of the nose and mouth. Usually, this clears the nasal passage of whatever was bothering it.

Your **respiratory system** has other responses that keep the airways clear. Coughing, yawning, and fainting all keep you breathing properly. Your body handles all of them automatically.

The Body's Dust Trapper

Before many dirt, dust, or other invading particles reach the lungs, they get trapped in a layer of sticky **mucus** that lines most of the airways. Beneath the mucus are thousands of microscopic hairs called **cilia.** The cilia move back and forth, sweeping the dirt and mucus away from the lungs. When mucus reaches the throat, it is either swallowed or expelled from the body by coughing. In the picture at right, a dirt particle *(brown)* is stuck in cilia.

Ah-choo!

When you sneeze, you spray a fine mist of mucus as far as 3 m (10 ft.) into the air. **Germs** and other particles fly out of your nose and mouth as fast as 166 km/h (103 mph).

In the longest recorded sneezing fit, a woman from England sneezed for 977 days. It's estimated that she sneezed one million times in the first year. That's a lot of tissues!

What's a Cough?

A cough is like a sneeze, but the irritation that triggers it starts in the throat, the windpipe, or the lungs, not the nose. When you cough, you take a deep breath, then tightly close your vocal cords and contract your chest muscles. Air pressure builds up in the lungs. Then you open your vocal cords and cough out both the air and the mucus.

The Big Squeeze

If your brain doesn't get enough oxygen, sometimes your body will briefly shut down, making you lose consciousness, or faint. This used to happen often to women in the late 19th and early 20th centuries, all because of a fashion trend.

Tiny waists were the rage in those days, so women wore an under-garment called a corset *(above, right).* Pulled tight, the laces in back of the corset squeezed the waist—and the lungs and **diaphragm** *(right).* Unable to breathe properly, women fainted, recovering once the laces were cut and they could breathe again.

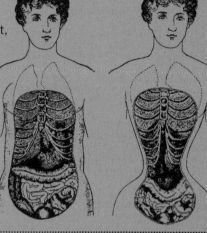

Without Corset **With Corset**

Why Do We Yawn?

When people—and animals, too—get tired or bored, some-times their breathing slows down and becomes shallow. This increases carbon dioxide in the blood and decreases oxygen. In response, the body takes an extra-deep breath, called a yawn. A yawn is your body's way of giving itself a quick dose of oxygen while at the same time getting rid of too much carbon dioxide.

What is the Nervous System?

The **nervous system** controls all of the body's activities, from walking and talking to seeing, hearing, and breathing. It is composed of three parts: the brain *(pages 50–51)*, the **spinal cord,** and **the nerves.** Together, the brain and the spinal cord are called the central nervous system (CNS), which is the command center of the body. The nerves that branch out from the brain and the spinal cord make up the peripheral nervous system (PNS). These nerves relay messages between the CNS and the rest of the body. Each nerve is made up of many **neurons,** or nerve **cells.** There are two main types of neurons—sensory and motor. Sensory neurons carry information about what you see, hear, taste, touch, or smell to the CNS to be processed. The CNS then tells the body how to respond to this information by sending commands to **muscles** and other parts of the body via the motor neurons.

Think Fast!

When a ball flies toward you, you have to think fast! And that's exactly what happens. The eyes see the ball coming and send this information to the brain, which immediately tells the muscles how to respond. This process takes only a split second—giving you plenty of time to hit the ball!

Sensory Input → **Processing** → **Motor Output**

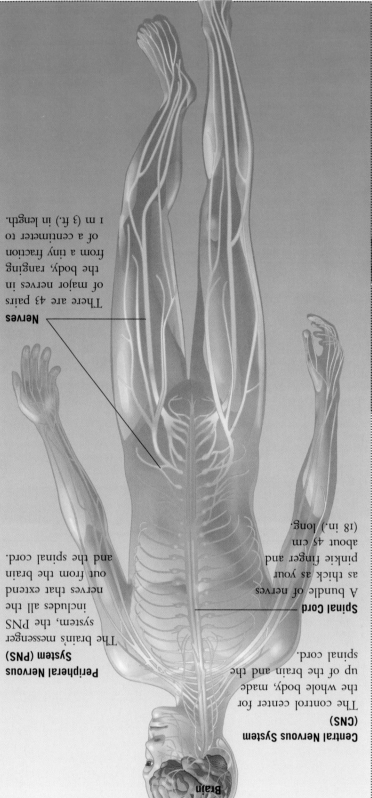

Central Nervous System (CNS)

The control center for the whole body, made up of the brain and the spinal cord.

Peripheral Nervous System (PNS)

The brain's messenger system, the PNS includes all the nerves that extend out from the brain and the spinal cord.

Spinal Cord

A bundle of nerves as thick as your pinkie finger and about 45 cm (18 in.) long.

Nerves

There are 43 pairs of major nerves in the body, ranging from a tiny fraction of a centimeter to 1 m (3 ft.) in length.

Brain

Let's Compare

Neurons and Nerves

A neuron (or nerve cell) looks like a bug with spidery legs and a long tail. The legs, called dendrites, receive electrical signals. These travel along the cell's tail—the axon—which is insulated by a layer of **fat** called the myelin sheath. When the signal reaches a synapse—a tiny gap between two neurons—it jumps from one neuron to the next and continues its journey.

Nerves are made up of bundles containing thousands of neurons. A cross section of a nerve (below) shows these bundles (pink circles).

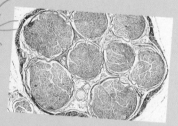

- Synapse
- Myelin Sheath
- Axon
- Dendrite
- Cell Body
- Nucleus

What's a Reflex?

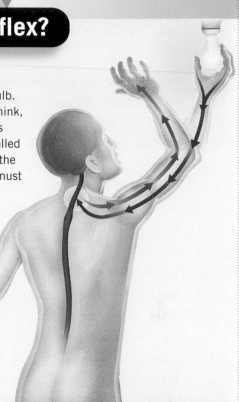

You touch a hot light bulb. Before you have time to think, your hand jerks away. This involuntary response is called a **reflex.** Reflexes protect the body from harm, so they must happen quickly. To speed up the response, neurons bypass the brain and send information directly to the spinal cord, which processes it immediately. The brain registers the pain milliseconds later. Ouch!

Mind over Body

How can a street performer like this man swallow a sword that is 0.6 m (2 ft.) long? He has learned to control his gag reflex—an involuntary contraction of the throat muscles that keeps you from swallowing something large, distasteful, or toxic. Some people have even learned to slow their breathing —another involuntary process—so that they require very little oxygen. This allows them to be buried alive for as long as 45 minutes. But these feats should be done only by professionals, so please don't try them at home!

Nerves Exposed!

No, this isn't a character from the latest horror movie, but it is a little creepy. You are looking at the nerves of Harriet Cole. They are all that remains of this 19th-century woman who donated her body to science after she died from tuberculosis at the age of 35. Harriet's employer, Dr. Rufus B. Weaver of Philadelphia's Hahnemann Medical College, made medical history in 1888 when he spent five months carefully dissecting, exposing, and then mounting Harriet's nerves. Her eyes, which have been preserved, remain attached to the **optic nerves.**

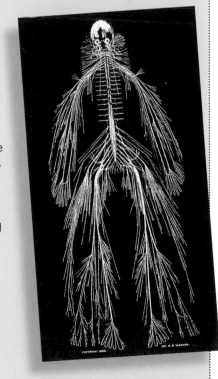

The Brain

Parts of the Brain

More complex than any computer, the brain is the control center for the entire body. It keeps us alive by making sure our heart beats and our lungs breathe. It also signals when the body needs fuel by making us feel hungry or thirsty, and it controls our digestion. But the most incredible part is that the brain gives each of us the ability to think, learn, create, and dream; to feel happy or sad; and to have hopes and fears. The brain is what makes each of us a unique individual.

At roughly the size of a grapefruit, the brain is pinkish gray and weighs about 1.4 kg (3 lb.). It is lumpy, and like a gelatin dessert it is squishy. Lots of blood vessels run through it. The brain is basically a mass of nerve **cells** called **neurons** *(page 49)*—about 100 billion—but it contains no sensory neurons, so it can't feel pain.

The brain has three main sections: the **brainstem,** the **cerebellum,** and the **cerebrum.** Each has a specific job to do, but they rely on one another to control the intricate workings of the human mind and body.

Cerebrum
This is the largest part of the brain. Its wrinkled outer layer, the **cerebral cortex,** is where thinking takes place.

Hypothalamus
This structure regulates body temperature, heart rate, hunger, and thirst, and with the **pituitary gland,** regulates growth and sexual development.

Brainstem
The brainstem regulates basic life functions such as heartbeat, breathing, swallowing, and waste elimination.

Cerebellum
The cerebellum governs balance, muscle movement, and posture.

Pituitary Gland
This gland regulates growth.

Protecting the Brain

By wearing a safety helmet, this skateboarder is really using his head! The brain has a lot of natural protection. It floats in a special liquid called cerebrospinal fluid, which acts as a shock absorber. Around this are three cushioning layers of **membranes** called meninges. Then there is your hard, bony skull. Your scalp and hair provide even more protection. But a fall or a hard tackle can still cause serious injury, so always wear a helmet.

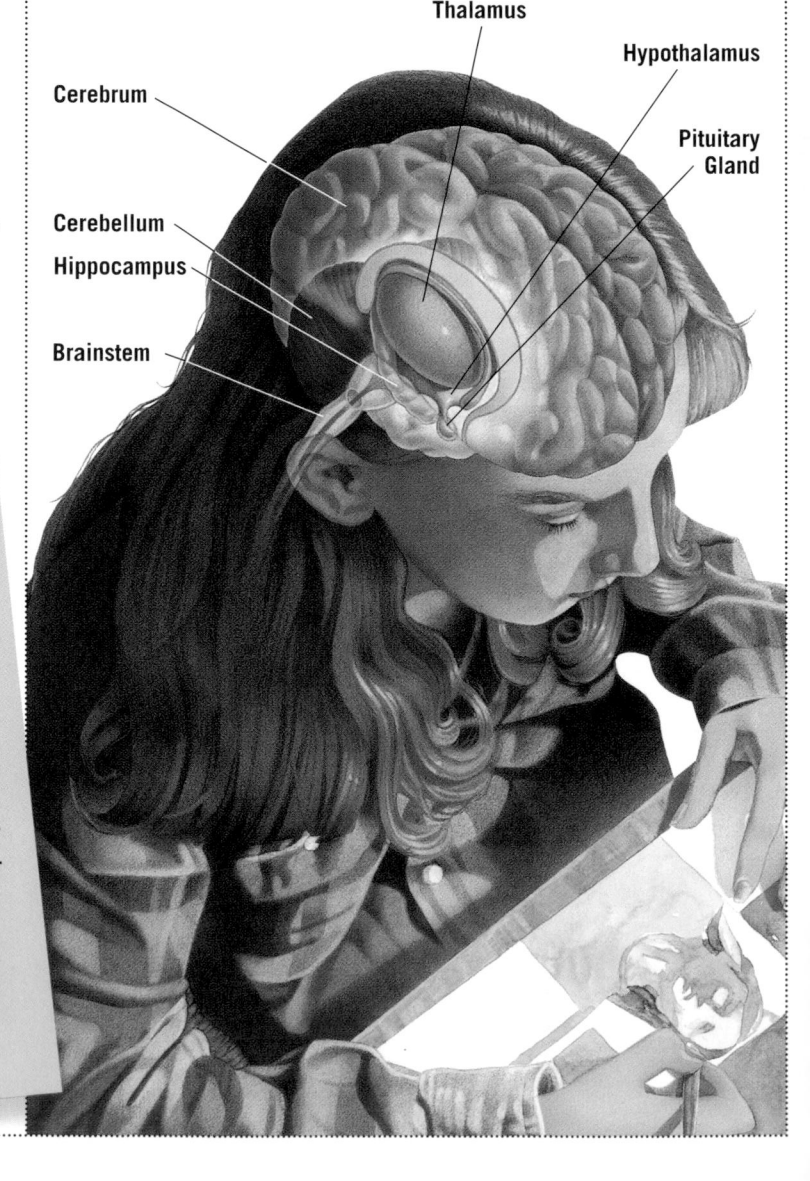

Thalamus
Hypothalamus
Cerebrum
Pituitary Gland
Cerebellum
Hippocampus
Brainstem

Blood to Brain

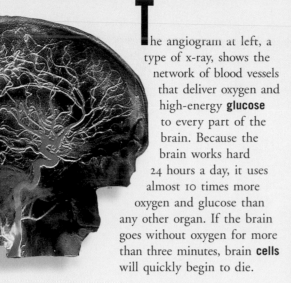

The angiogram at left, a type of x-ray, shows the network of blood vessels that deliver oxygen and high-energy **glucose** to every part of the brain. Because the brain works hard 24 hours a day, it uses almost 10 times more oxygen and glucose than any other organ. If the brain goes without oxygen for more than three minutes, brain **cells** will quickly begin to die.

What's Happening in the Brain

Imagine being able to "see" inside the brain without having to open up the skull. Special computer-driven machines allow just that, helping scientists and doctors not only figure out what's wrong with a damaged brain, but also "see" a healthy brain in action.

CAT Scan

A CAT (computerized axial tomography) scan combines x-rays with computer technology to produce a precise image of a cross section of the brain. It is used to detect tumors, blood clots, birth defects, and certain kinds of brain damage.

Let's Compare

Brain Sizes

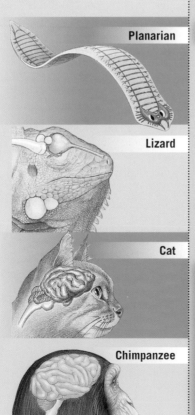

Planarian

Lizard

Cat

Chimpanzee

Usually, the bigger the animal, the bigger the brain. But brain size alone does not determine intelligence. The size and complexity of different parts of the brain are what count. For example, a flatworm called a planarian has the simplest brain: a cluster of nerve tissue beneath the eyespots. A lizard's brain is more complex but still small; it functions mainly to keep the reptile alive. Mammals, such as cats and chimpanzees, have a brain that is much more complex, especially the wrinkled cerebral cortex—the area of learning and intelligence. But their brain is still not as large or as complex as that of humans. The human brain is larger in relation to body size than the brain of any other animal.

PET Scan

A PET (positron emission tomography) scan is a picture of the brain at work. When the brain works it uses glucose. A PET scan shows glucose levels in the brain. Shades of red and yellow indicate the most active regions. Areas that are green and blue are less active.

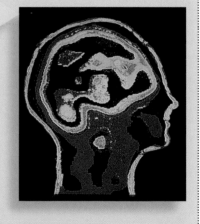

MRI

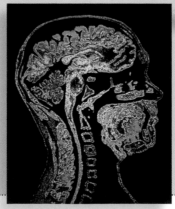

MRI (magnetic resonance imaging) reveals blood vessels, **nerves**, brain, **spinal cord**, and other types of soft **tissue**. It uses a strong but harmless magnetic field to produce detailed cross-section images that can show abnormalities and brain diseases.

Automatic Brain

Two parts of the brain—the **brainstem** and the **cerebellum**—are devoted entirely to keeping us alive. They do this automatically, without our consciously thinking about it.

The brainstem, which is the oldest section of the brain in evolutionary terms, is made up of the medulla oblongata, the pons, and the midbrain. It carries information between the body and the brain, and it also regulates such vital functions as breathing, **blood pressure,** swallowing, and sleep cycles.

At the top of the brainstem is the thalamus, the relay center for all our senses except smell, and the **hypothalamus,** which controls emotions, sleep, and feelings of hunger and thirst. Located behind the brainstem is the cerebellum. This part of the brain adjusts posture and balance and coordinates **muscle** movement.

Where in the Brain?

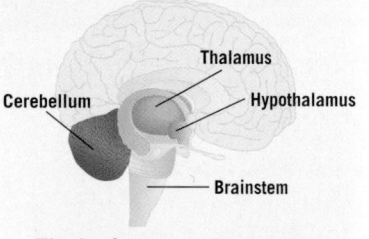

- Thalamus
- Cerebellum
- Hypothalamus
- Brainstem

The brainstem and cerebellum control the automatic functions of the body.

Every second, the brain receives millions of messages from all over the body. Luckily, a filter called the reticular formation screens out useless data—about 99 percent of the messages—so you aren't bombarded with information. For example, when you are really focused on a TV program or on doing homework, you probably won't be distracted by a dog barking or someone vacuuming. But this part of your brain can sometimes be overactive, and then you have trouble falling asleep at night. And have you ever had that sick feeling in your stomach when you remember something embarrassing? Well, that's your reticular formation at work again.

Cerebellum: A Balancing Act

Phantom Pain

People who have had an arm or leg amputated sometimes feel discomfort that seems to be coming from that limb. This so-called phantom pain occurs because there is a specific area in the brain assigned to each limb. When another body part, controlled by a nearby area of the brain, is touched, the missing limb's memory area may be stimulated.

The Busy Hypothalamus

Nestled deep within the brain is the hypothalamus, a small, cherry-size structure with a big job: to control many of the body's automatic physical and emotional functions.

First, the hypothalamus acts as a thermostat to hold the body's temperature at a constant 37°C (98.6°F). It also regulates feelings of hunger and thirst, and the state of being alert or sleepy. And it helps the pituitary gland trigger functions such as growth and sexual development. In addition, many strong emotions —fear, anger, happiness, and pleasure—arise from the hypothalamus.

Emotions

Temperature Control

Sexual Development

Baby's Smile

Nothing is cuter than the smile of a new baby. And there is a very good reason for this: That adorable smile is an instinct—a behavior that is not learned but is already in the brain at birth. Through this engaging behavior, the baby encourages its parents to respond—by talking to the baby, playing with it, and holding it. This attention stimulates the development of the baby's **nervous system**. It is necessary for healthy mental and physical development.

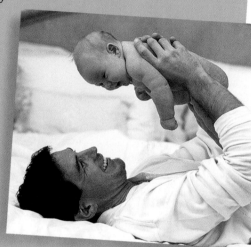

Balancing safely on a tight-rope takes years of practice. It also requires the teamwork of two parts of the brain: The **cerebrum** tells the muscles to move, and the cerebellum automatically coordinates fine motor skills such as balance and posture to ensure smooth movement.

The cerebellum stores up memories of previously learned movements in its two mounds of deeply folded **tissue**. In the case of a tightrope walk, for example, the cerebellum would call on those memories to tell the body how to position itself.

Information to and from other parts of the brain pass through the cerebellum along a tree-shaped branching **nerve** pathway called the arbor vitae, Latin for "tree of life" (below).

Looking more like a cauliflower than a part of the brain, a cross section of the cerebellum reveals the arbor vitae, a branching nerve pathway.

Conditioned Reflex

Early in the 20th century, Russian physiologist Ivan Pavlov wanted to test a theory. He knew that when a dog smelled food, it would salivate—a normal, involuntary **reflex.** But Pavlov wanted to know whether a dog could be taught to salivate when given a stimulus other than food—in other words, if it could be taught a reflex.

As part of his experiment, Pavlov rang a bell just before he fed a hungry dog some meat. Soon the dog associated the sound of the bell with food and began to salivate at the sound alone. The white-bearded Pavlov—shown here with his assistants and a dog—had proved that inborn behavior could be shaped by learning, creating what is known as a conditioned reflex.

Emotions

Why do we laugh at a joke, cry when we watch a sad movie, or get angry when someone hurts us? These emotions are all formed in a cluster of structures called the limbic **system,** located in the center of the brain between the **brainstem** and **cerebrum.** This location is next to the thinking part of the brain, which suggests that the limbic system may mix reason and emotion.

The limbic system includes the **hypothalamus,** the hippocampus, and the amygdala. The amygdala determines the appropriate behavior for a particular feeling—crying when you are sad or laughing when you are happy. The hypothalamus triggers the physical feelings you have with different emotions such as a pounding heart when you are afraid. And the hippocampus makes sure you remember emotional incidents.

Where in the Brain?

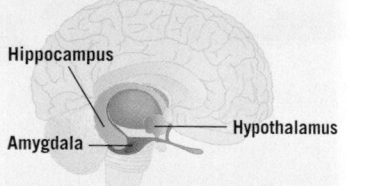

Hippocampus

Amygdala — Hypothalamus

The limbic system controls emotions and is located in the middle of the brain.

Happy or Sad

Laughing until you cry sounds silly, but in fact it often happens. We all know that joy, sorrow, and frustration can bring on tears. But sad or uncomfortable situations can sometimes make you suddenly start to giggle. This happens because laughing and crying involve many of the same **muscles** and brain circuits.

The Humors

People in the Middle Ages thought that emotions and personality were shaped by four different body fluids—blood, phlegm, black bile, and yellow bile. They called these fluids "humors." Someone who was cheerful and optimistic (far right) was thought to have an abundance of blood, whereas someone indifferent and unemotional (near right) had too much phlegm.

Sadness was thought to come from too much black bile, and an overflow of yellow bile made you irritable and angry. Today, we still refer to someone as good- or bad-humored when we describe that person's mood or temperament.

What Triggers Emotions

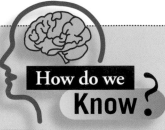

He may not be smiling, but this is one happy rat! An electrode has been embedded in the part of its hypothalamus called the pleasure center, and each time the rat taps the lever, a small current of electricity briefly stimulates this center. In fact, it feels so good that the rat sometimes taps up to 5,000 times an hour! Experiments like this one—which doesn't hurt the rat at all—help scientists better understand how the brain works.

Just Say No!

Some people think it's cool to smoke cigarettes, drink alcohol, or take drugs, but it's actually pretty stupid. The chemicals in these products affect the brain, usually interfering with how messages travel from one **neuron** *(page 49)* to another. By artificially creating a heightened sense of alertness, relaxation, or pleasure, they set the stage for an addiction: Either the person believes he or she needs the drug to feel good, or the body believes it needs the drug to function normally. How do you avoid getting hooked? Just say no!

A Bumpy Map of the Brain

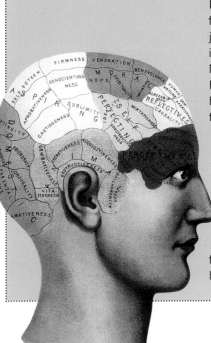

Do you think you could figure out a friend's personality just by feeling the bumps and indentations on his or her skull? Not likely, but Viennese doctor Franz Joseph Gall thought so back in the 1790s. He divided the skull's surface into more than 100 areas *(left)*, claiming that each one corresponded to a specific personality trait. He called his system phrenology. For example, he thought a bump in front of the ear meant a person was good at paying attention. But a bump over the right ear meant the person had destructive tendencies.

What's Hypnosis?

Although people who are hypnotized appear to be asleep, they are actually just deeply relaxed. The brain is alert and very susceptible to suggestion, allowing people such as this Victorian lady to perform feats they could never do otherwise.

The 18th-century physician Franz Anton Mesmer thought hypnosis could cure some nervous ailments. Today, it is used to help people quit smoking, lose weight, or get rid of warts. It is sometimes used to anesthetize patients during surgery.

Thinking Brain

Looking like a wrinkled walnut, the two halves, or hemispheres, of the **cerebrum** make up 85 percent of the brain's mass. Its surface, called the **cerebral cortex,** is where we think, make decisions, learn, interpret sensations, and store memories. It is what we call the mind, and is the part of the brain that truly sets humans apart from any other animal.

The cortex is shiny and pinkish gray in color. Although it is only 3 mm (0.12 in.) thick, the wrinkled surface contains a lot more brain **cells** than if it were smooth. Think of it this way: If you were to wad up a sheet of newspaper into a ball the size of your fist, it would have the same number of words in a much smaller space. This is true of your cerebral cortex, which can hold a lot more **nerve** cells—about 50 billion—in a much smaller space. If your cortex were smooth, your head would have to be three times larger to have the same thinking power.

Speech · Creativity · Mathematics · Pattern Recognition · Right Hand · Left Hand · Science · Left · Right · Judging Distance · Writing and Language · Musical Ability · Problem Solving · Imagination

Gorilla to the Rescue

On a summer day in 1996, Binti Jua, a gorilla at Chicago's Brookfield Zoo, came to the rescue of a three-year-old boy who had fallen 5.5 m (18 ft.) into the gorilla enclosure. Binti checked the unconscious toddler twice for signs of life. Then she carefully picked him up and carried him toward the enclosure door, where she gently laid the boy down. Whether it was instinct or conscious thought on Binti's part, bystanders were awed by the gorilla's actions and her compassion.

Left Brain, Right Brain

The left and right hemispheres of the brain look alike and work together, but each side is in charge of different activities. The left side typically controls logical thinking and language, for example, whereas the right side dominates in creativity. The two sides communicate through the corpus callosum, a thick bridge composed of millions of neurons. In the **brainstem,** the **nerves** from the left and right hemispheres change sides. As a result, nerves in the left hemisphere control the right side of the body, and vice versa. And because the left hemisphere controls language, most people write with their right hand. In left-handed people, the right side of the brain controls language.

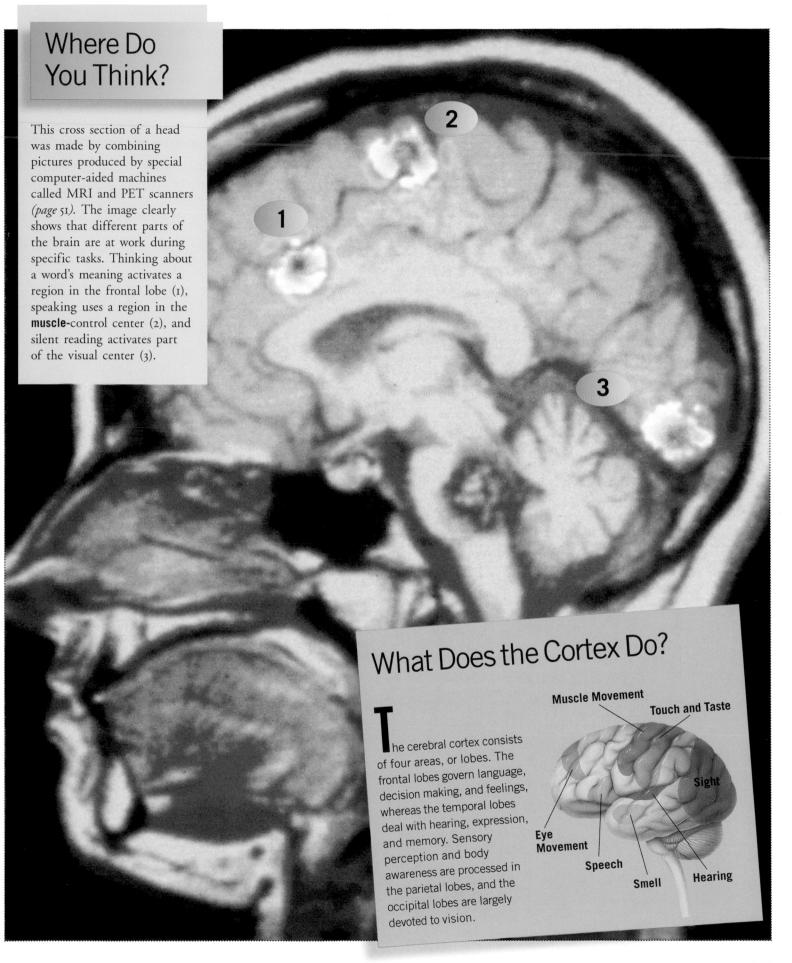

Where Do You Think?

This cross section of a head was made by combining pictures produced by special computer-aided machines called MRI and PET scanners *(page 51)*. The image clearly shows that different parts of the brain are at work during specific tasks. Thinking about a word's meaning activates a region in the frontal lobe (1), speaking uses a region in the **muscle**-control center (2), and silent reading activates part of the visual center (3).

What Does the Cortex Do?

The cerebral cortex consists of four areas, or lobes. The frontal lobes govern language, decision making, and feelings, whereas the temporal lobes deal with hearing, expression, and memory. Sensory perception and body awareness are processed in the parietal lobes, and the occipital lobes are largely devoted to vision.

Muscle Movement
Touch and Taste
Sight
Eye Movement
Speech
Smell
Hearing

Learning Brain   How We Learn

Throughout our lives, we learn many things: how to walk, talk, behave, ride a bike, read, write, and make friends, to name only a few. This process occurs in the **cerebral cortex,** the area of intelligence and understanding. Any kind of learning involves memory, or the ability to recall something. In physical terms, a memory is actually a specific pathway that the **nerve** impulses follow through the network of the brain's billions of **neurons.**

Although many parts of the brain work together to create, store, and call up memories, the process of filing away a memory involves areas such as the temporal lobe and the hippocampus, which helps change short-term memories into long-term memories. The amygdala helps store emotional memories, especially smell.

Where in the Brain?

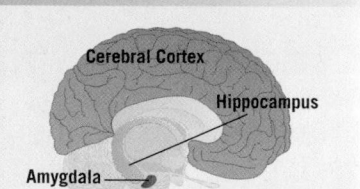

Cerebral Cortex

Hippocampus

Amygdala

Learning takes place in the cerebral cortex, the outer layer of the cerebrum.

Writing is probably easy for you, but it can take a long time for a young child to learn. First there is scribbling with a crayon, then holding a pencil properly, then writing a word, and finally writing a sentence. Learning involves remembering both mental and physical skills that are developed and fine-tuned over a long period of time.

A toddler's drawing *(top)* shows big hand movements used to make marks on paper. Soon the scribbles are controlled *(middle),* and there is an understanding that certain marks are used in writing. The next step *(third from top)* is knowing and forming the letters of the alphabet, and putting them together to make nonsensical words. And finally, the payoff: You can read and write *(bottom)!*

Mirror Image

Would **You** *Believe?*

Hey, that's me in the mirror! This toddler has only recently realized that he is seeing his reflection and not another child. Like learning to walk, it takes time to learn to recognize yourself in a mirror or a photograph. Chimpanzees seem to be the only other animal with any level of self-recognition.

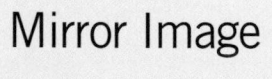

Neuron Connections

A baby is born with about 100 billion neurons—the most the brain will ever have. At first they are widely spaced *(top),* but as the baby learns, connections begin to grow between them. By the time the child is about two years old, each neuron has some 15,000 connections.

We continue to learn and make new connections. But from age 20 onward the brain shrinks, losing thousands of brain cells a year. During a lifetime, people lose about 10 percent of their brain cells.

Infant

Toddler

Elderly Adult

Short- and Long-Term Memory

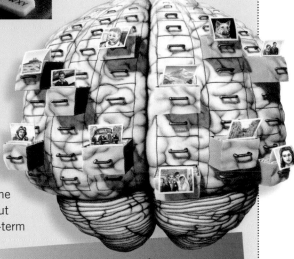

memories—such as a joke you tell often or what happened on your first day of school—are filed away by your brain and can last forever.

W hat can hold 1,000 times the information contained in a 20-volume encyclopedia? Your memory! Short-term memories, such as an unfamiliar telephone number, last only about 30 seconds. But long-term

Memory Helpers

Can't remember which months have 30 days and which have 31? Your knuckles can help! Starting with the pinkie knuckle of the left hand, name a month on each knuckle and on the spaces in between *(below).* The knuckle months have 31 days and the rest have 30 days.

Easy, right? If you have

something to memorize, try making up a rhyme, jingle, abbreviation, or visual clue. For example, the name Roy G. Biv (from the first letters of the words "red," "orange," "yellow," "green," "blue," "indigo," "violet") is a way to recall the order of the colors in a rainbow.

January	March	May	July
February	April	June	

August	October	December
September	November	

People   Dutch Calculator

S ome people can do amazing things with numbers. One such person was Antoon Van den Murk, a Dutch farmer who became famous in the 1950s for his ability to crunch numbers in his head. The Dutch calculator explained how he multiplied such large numbers. He used 6,341,082,426 times 38,254,319,074 as an example. "You just multiply 6,341,082,426 by 38,000,000,000 and remember the solution. Then you multiply 6,341,082,426 by 254,000,000, by 319,000, and by 74. Add them up in your head and you have the answer." Now you try it! Van

den Murk's talents did not bring him much success later on. He eventually settled down to work as a laboratory assistant.

What Is Sleep?

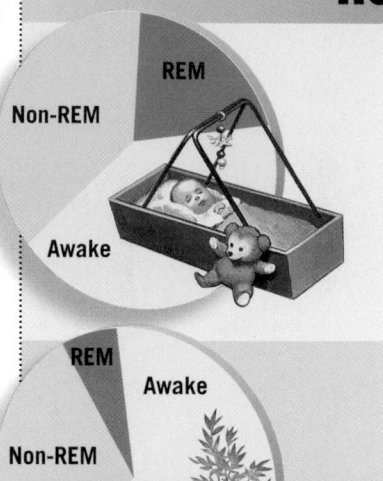

Sleep is a time for both the body and the brain to rest. While our body functions obviously don't turn themselves off completely, **muscles** relax, breathing and heart rate slow down, and **blood pressure** *(page 28)* drops. When you wake up in the morning, you feel refreshed. Lack of sleep can have the opposite effect: After a bad night's sleep, you will feel tired and cranky. Two nights of bad sleep will make your mind feel fuzzy, and if you don't get any sleep for five days, you will start hallucinating—seeing things that don't really exist.

Scientists have identified two types of sleep: REM—for rapid eye movement—sleep, during which intense dreaming occurs, and a peaceful period called non-REM sleep. During REM sleep the eyes dart rapidly back and forth under the closed lids, almost as if the sleeping person were watching something. It is during this period that sleep experts believe the brain is organizing the events of the day.

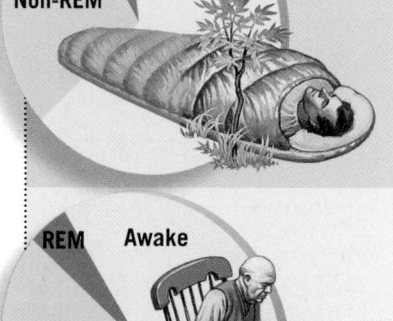

Baby

A baby needs about 14 hours of sleep a day, a large part of which is spent in REM sleep. The longer REM sleep is thought to help create essential **neuron** connections.

Adult

An adult needs about seven to eight hours of sleep a night, 20 percent of which is spent in REM sleep.

Elderly Adult

By age 70, a person needs only about six hours of sleep, but about 20 percent is still spent in REM sleep.

Stages of Sleep

Every night, you cycle through several stages, or levels, of sleep, each lasting about 90 minutes. As you drift off (Stage 1) your body relaxes, and for a few minutes your mind wanders with thoughts and minidreams. You then fall asleep (Stage 2) and gradually move into deep, or non-REM, sleep (Stages 3 and 4). Your mind is resting and quiet.

At the end of Stage 4 sleep, you usually change body position, and your brain suddenly becomes very active, with lots of vivid dreaming. This is REM, or shallow, sleep. After about 20 minutes, the brain drifts back to another round of non-REM sleep. This cycle repeats itself about five times during the night. Toward morning, you sleep more lightly, and then awaken.

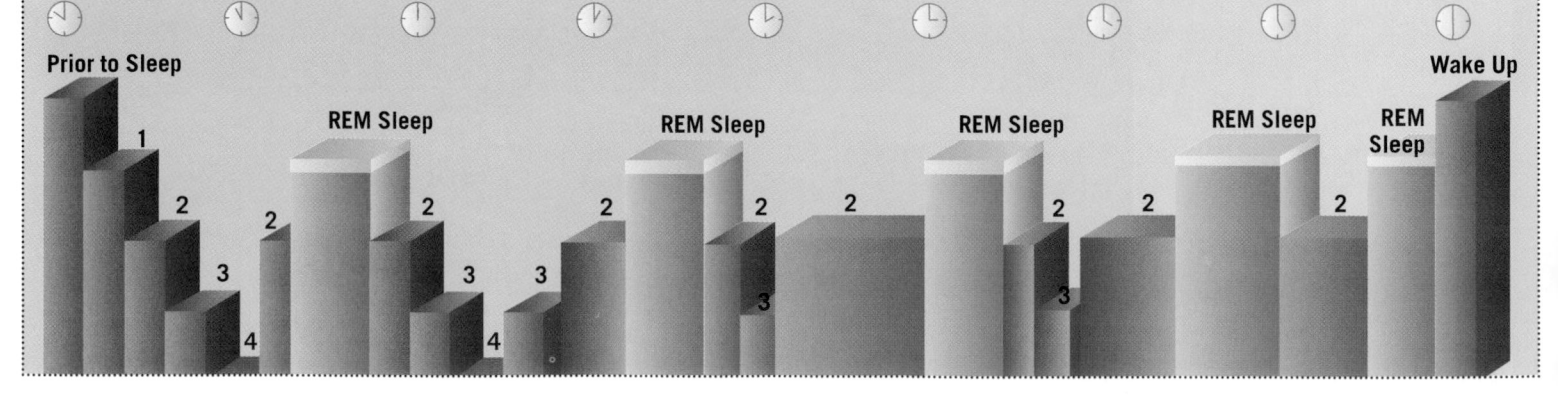

What's it Like?

Living Underground

For six months in 1972, French geologist Michel Siffre lived underground, in a Texas cave, with no way of knowing what time it was. Siffre was helping scientists study the body's 24-hour sleeping and waking cycle, called the circadian rhythm. The rhythm is regulated by the earth's daily rotation, but even without time cues, Siffre's body created its own cycle, averaging about a 28-hour "day."

"Falling" Asleep

Have you ever been drifting peacefully off to sleep when suddenly you feel like you are falling? Sometimes you may see yourself in a dream falling or stepping off something. You wake with a jerk. Upon awakening, you are surprised to discover that you weren't falling at all; you just dreamed it.

Scientists call this falling sensation a "hypnic jerk" or a "nap jerk." Although they have given the sensation a name, they are not sure what causes it. They do know, however, that it occurs during Stage 1, non-REM sleep (*diagram, left*).

Let's Compare

Sleep versus Hibernation

To escape cold weather and no food, this ground squirrel will hibernate through the long winter to conserve energy. During hibernation, heart and breathing rates slow and body temperature drops to near freezing.

Sleep, though, lasts only an average of seven to eight hours. Heart and breathing rates slow during non-REM sleep but quicken during active REM sleep. And the body's temperature remains pretty much the same.

Sleep Deprivation

In 1964, 17-year-old Randy Gardener became the subject of his own science fair project on sleep deprivation. Monitored by doctors and sleep experts, the high-

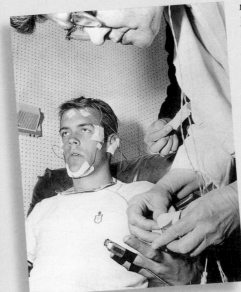

school senior aimed to stay awake for a record-breaking 264 hours (11 days). As the days passed, Randy gradually lost eye focus and hand co-ordination and had memory lapses. Then he started to hallucinate, thinking a street sign was a person. In the end, he officially made *The Guinness Book of World Records* and celebrated by sleeping for 14½ hours.

Dreaming

Every night we slip into a world as full of adventure, detailed imagery, and strong emotions as an action movie. These vivid dreams occur most often during REM sleep *(pages 60-61)*, and at least one dream occurs every REM period. They may be short at first—three to five minutes long—but as the night goes on, dreams can last up to 45 minutes.

Dreams that occur in the deep, non-REM sleep tend to be about everyday experiences, like playing with a dog. You may even act out these dreams by walking or talking in your sleep *(opposite)*.

No one knows for sure why we dream, but scientists have some theories. Some believe the brain is sorting out the day's events and storing memories. Others think dreams are a way of telling us about our hidden wishes and fears.

The Realm of Hypnos

The sculpture fragment above shows Hypnos, the ancient Greek god of sleep, who was said to lead mortals to slumber by fanning them with his wings. The son of Nyx, the god of night, Hypnos was also the father of Morpheus, who was the god of dreams and believed to be the brother of figures who appeared in dreams. The ancient Greeks believed that dreams were messages from the gods.

Common Dream Themes

Everyone dreams, but not everyone remembers their dreams. Although no one knows for sure, some scientists think this is because your memory—like your body—may be resting during sleep.

The most fantastic dreams occur during REM sleep, when the brain is very active. During this part of the sleep cycle, your dreams may be in color and have sound. Time is usually distorted. Feelings and emotions from the day may invade your dreams. People who do remember their dreams often tell about flying or falling through the air, or being chased; encountering frightening animals; or seeing fire, water, snakes, or horses. These images come up over and over in dreams. But even with all of the action going on, the brain keeps the body almost paralyzed so you won't physically act out your dreams and hurt yourself.

Brain Waves

Electrodes placed on the head can detect the brain's natural electrical activity. Connected to an electroencephalograph, or EEG *(right)*, this activity appears as wavy lines, or brain waves. Brain waves look different depending on what a person is doing.

Alpha Waves

Alpha waves occur when a person is awake but deeply relaxed, with eyes closed.

Beta Waves

Beta waves occur when a person is awake and alert. This type of brain wave is present during most mental activities.

Theta Waves

These brain waves are found mostly in children. They rarely occur in adults except just before waking from sleep.

Delta Waves

Delta waves occur when an adult is in deep sleep, corresponding to Stages 3 and 4 of the sleep cycle *(pages 60-61)*.

Sleepwalking

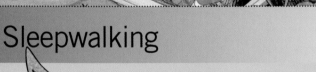

Bugs Bunny must have gotten thirsty in his sleep. Sometimes the brain can't stop commands from going to the **muscles.** When this happens, people walk and talk while sleeping. They may act out a dream—or, like Bugs, get a drink of water. Even though their eyes are open, sleepwalkers are in deep, non-REM sleep. The next morning, they don't remember anything about their nighttime roamings.

Do Animals Dream?

A litter of dozing puppies lie in a jumbled heap. Are they dreaming? Probably. The area of the brain that controls sleep —the **brainstem**—is similar in all vertebrates (animals with backbones). However, only mammals are known to dream. According to laboratory tests on a wide range of mammals, all of the creatures exhibited both REM and non-REM sleep. The only mammal tested that didn't was the short-beaked echidna from Australia.

Endocrine System

T he **nervous system** is essential to sustaining life, but it does not work alone. It needs the help of the **endocrine system**, a network of **glands** and **organs** scattered throughout the body.

The endocrine system affects the body over a period of days, weeks, or even years. Using chemical messengers called **hormones**, it influences such functions as growth, sexual development, **nutrient** balance in the blood, and how the body reacts to stress.

The major organs of the endocrine system include the adrenal, thyroid, parathyroid, pineal, pituitary, and thymus glands as well as the pancreas, **ovaries** in women, and testes in men. The **hypothalamus**, which is really part of the nervous system, is also considered an endocrine organ because it produces several hormones. The **pituitary gland** in the brain regulates the activities of all the other endocrine glands.

What's a Hormone?

The pattern below is not some-thing you'd see through a kaleidoscope; it's the crystalline form of the hormone esterol, which is one of the female hormones. Produced mainly by the endocrine glands, hormones are chemical messengers that can either act on one organ or travel through the blood-stream to affect a number of organs.

These messengers help control your body's activities, such as growth, sexual development, and **metabolism.**

Testes
Produce the male hormones that give men their male characteristics.

Ovaries
Produce female hormones that give women their female characteristics.

Pancreas
Regulates levels of blood sugar.

and regulate metabolism and development of sexual traits.

Thymus Gland
Helps in the development of antibodies, important to the immune system.

Adrenal Glands
Influencing the body in many ways, these glands help you react to stress, keep body fluids in balance,

Parathyroid Glands
Maintain proper calcium levels in the blood.

Thyroid Gland
Regulates metabolism and energy level.

Pituitary Gland
This master gland controls all other hormone-producing organs.

Pineal Gland
Influences sleep and growth of reproductive organs.

Hypothalamus
A structure in the brain that links the nervous and endocrine systems.

Fight or Flight

The terrified baboon at right bares its teeth as a leopard goes in for the kill. When faced with a dangerous situation, animals and humans experience what is called the fight-or-flight response. Hormones released by the adrenal glands speed up heart rate, raise **blood pressure,** quicken breathing, and shut down digestion, all in preparation for combat or escape.

Scaring Ourselves Silly

Looping upside down on a roller coaster, bungee jumping off a bridge, and watching horror movies in the dark are some of the many ways we love to scare ourselves. As we experience the sensation of fright, the adrenal glands begin to pump hormones that generate the heart-pounding fight-or-flight response. But the logical brain knows that these activities most likely won't harm us. The relief that knowledge brings turns a scary situation into a thrill.

Growing Up

Becoming taller and developing sexually are part of growing up *(pages 118-119)*. Everyone goes through this stage, called **puberty,** but each individual grows and develops at his or her own rate. It all has to do with timing—when growth and sex hormones are released.

Take these sextuplets, for example. They are all the same age, grew up in the same house, and had the same nutrition and care, yet each one is at a different stage of growth. Eventually, the late bloomers will naturally catch up with their speedy siblings.

Skin The Body's Container

S kin is the largest human **organ.** This outermost layer of your body makes up about 15 percent of your total weight. Skin has many purposes. It protects the body from injury and acts as the first line of defense against invading dirt and **germs.** Its sensitivity warns you when something is too hot or too sharp to touch. Skin also controls your body temperature. It even keeps your internal organs from drying out!

Skin is thin, but it is complex. It is made up of three layers: the outer **epidermis,** the dermis below that, and an underlying layer called subcutaneous **tissue.** The outermost **cells** that make up the epidermis are constantly dying, dropping off, and being replaced by new cells. The dermis, which is thicker than the epidermis, is strong and stretchy. It holds the **nerves** that allow us to sense pressure, pain, and temperature. Underneath it all is the subcutaneous layer, which contains **connective tissue** and **fat.**

Skin Deep

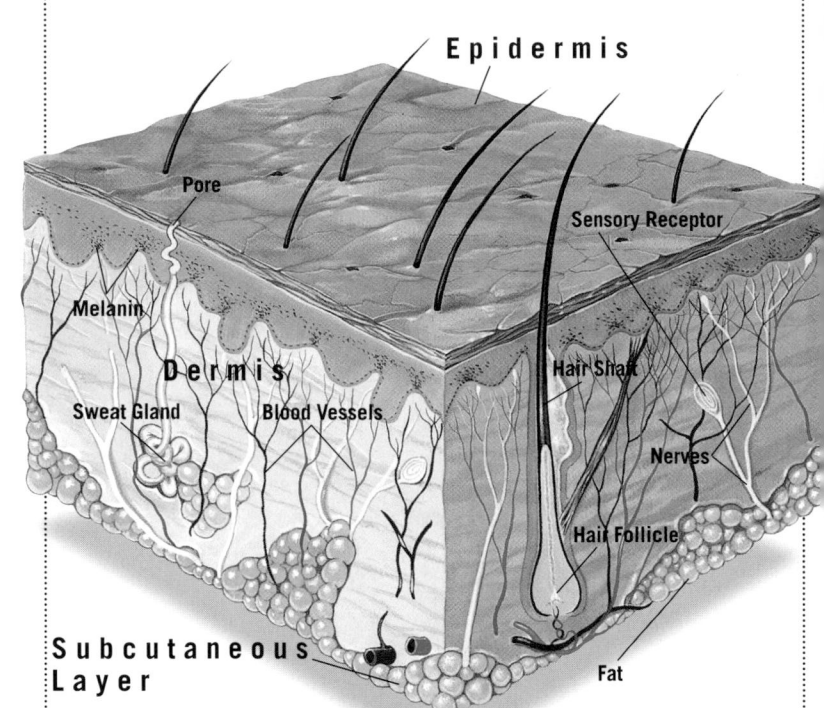

A cross section of skin reveals its paper-thin outer layer, the epidermis; the thicker dermis, laced with nerves and blood vessels; and its bottom layer of subcutaneous tissue padded with fat cells to protect the body's inner organs.

Let's Compare People of Color

A ll skin color comes from melanin, little grains of dark **pigment** in the epidermis *(diagrams below).* Melanin tints the skin in shades from pale tan to brown black. People with albinism, however, have little or no melanin in their hair, skin, or eyes, so they lack the coloring most people have. The skin color of different groups of people probably evolved in the climates they lived in long ago. Melanin protects skin from sun damage, so in hot countries, people developed more melanin in their skin.

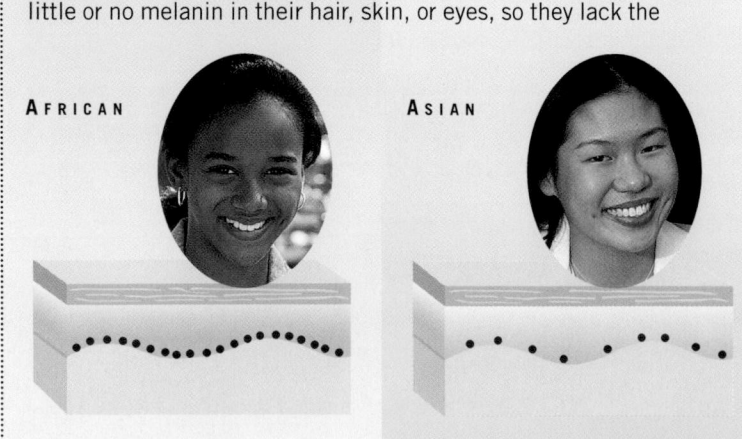

AFRICAN

ASIAN

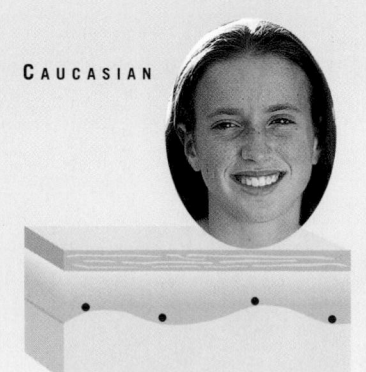

CAUCASIAN

CAUCASIAN WITH ALBINISM

Fingerprints: No Two Alike

The swirling ridges and grooves on each person's fingertips and toes are unique. No two people —not even identical twins— share the same patterns. No one has exactly the same pattern, but all patterns fall into one of three main types *(right)*. Which one do you have?

When a person touches something, signature fingerprints are left behind. Police officers have used fingerprinting to help identify and convict criminals since 1901.

Fingerprint ridges last a long time. When a 2,000-year-old Egyptian mummy was fingerprinted, its ridges were still perfectly visible!

LOOP PATTERN

WHORL PATTERN

ARCH PATTERN

Would **You** *Believe?*

Artificial Skin

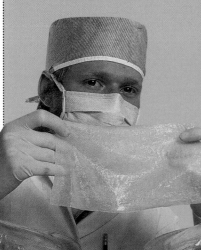

When a person is very badly burned, doctors sometimes need to cover the injuries with new skin in a procedure called a skin graft. In the past only a graft of a person's own skin would work. Today, however, skin can be made in the laboratory *(left)*. Artificial skin is made by growing skin cells on a kind of fabric that dissolves when it is put over a wound. This man-made skin provides a greater supply of new skin for burn victims so they can heal more quickly.

Put On Your Sunscreen!

Can you guess which of these people is older? Maybe it's not who you think! The monk on the left, who spent his life indoors, is 90. The Plains Indian woman on the right, who lived out in the blazing sun, is 56. Ultraviolet rays from the sun damaged cells in her epidermis, causing burning and wrinkling called "photoaging." Too much sun exposure can also make basic changes in cell structure, leading to skin **cancer**. Ultraviolet rays go through clouds and are present on cool days as well as warm ones. When you go outside, slather on that skin-protecting sunscreen.

Strange! But TRUE!

Mighty Mites

The monstrous-looking creature at right is really a tiny insect—a dust mite, magnified 900 times. Dust mites eat flaked-off bits of human skin. There's plenty to go around! In just one minute, 30,000 to 40,000 microscopic cells fall off your body as it rubs against clothes, sheets, and other objects. Don't worry, your body is continuously making new cells to replace the dead ones. Meanwhile, mites are crawling around your sheets, eyebrows, and eyelashes hungrily looking for their next meal.

Hair & Nails

More than three million hairs grow on a human body. There are only a few places where you won't find hair, such as your lips, the palms of your hands, and the soles of your feet. Most body hair is so fine it's barely visible. Hair, fingernails, and toenails are made of a tough **protein** called keratin and are rooted in the skin.

Hair and nails aren't there just to decorate our body. They have important jobs to do. Hair protects the skin. Pressure on the living roots of the hair helps us sense things that get close. Eyebrows and eyelashes keep dirt out of the eyes. The nails protect the sensitive ends of fingers and toes.

Have you ever wondered why it doesn't hurt to get your hair cut? Although the roots are alive, the shaft of the hair is made of **cells** that are dead and can't feel a thing. The visible part of nails is also made of dead cells.

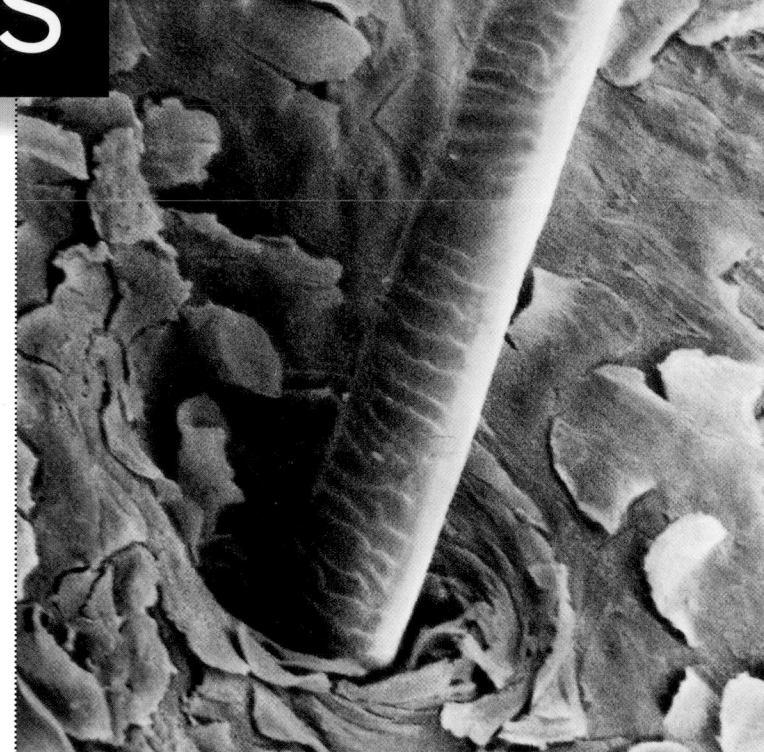

This photograph taken through a scanning electron microscope shows a hair rising from a hole, called a follicle, in the skin. Only the base of the hair is alive; its shaft is made of dead cells.

The Fingernail

A fingernail grows from a fold in the skin called the nail root. The visible part, called the nail plate, is made of dead cells. The nail plate is attached to the skin under it, called the nail bed, providing tough armor for the fingertip. Fingernails and toenails are embedded in sensitive **tissue,** so they act as antennae to convey touch to the fingers and toes.

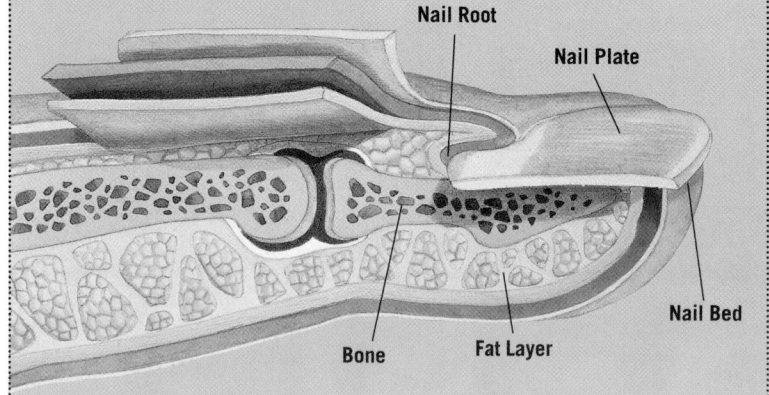

Nail Root

Nail Plate

Nail Bed

Bone

Fat Layer

Straight, Wavy, Curly

Is your hair naturally curly? Straight as a stick? Wavy? The shape of your hair follicles determines what kind of natural hairdo you have. Straight hair grows from a round follicle, and each hair is round. Wavy hair grows from an oval follicle; each hair is oval. Flat follicles coil flat hairs into curls. If you look at various kinds of hair under a microscope, you will be able to see these three different shapes.

Round Follicle
Straight Hair

Oval Follicle
Wavy Hair

Flat Follicle
Curly Hair

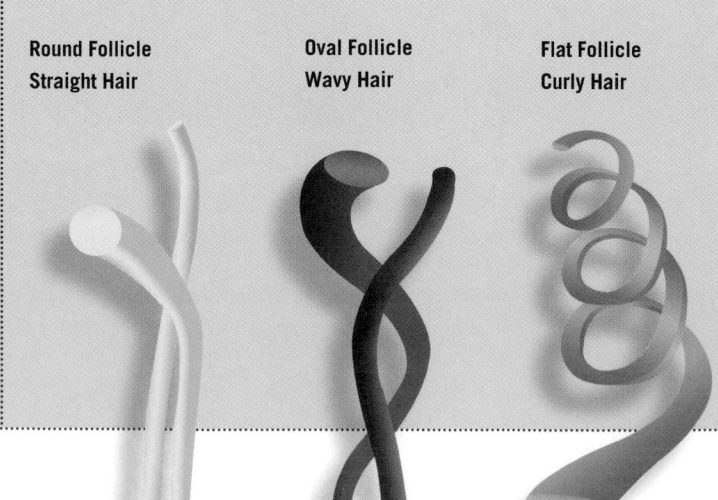

What are Goose Bumps?

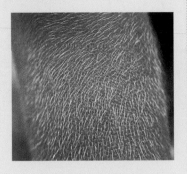

When you are cold, or when something startles you, you may feel your skin prickle as it becomes covered with little bumps. These are known as goosebumps because they look like the skin of a

Scaredy-Cat

plucked goose. Each bump forms when a tiny **muscle** next to a hair follicle contracts, pulling the follicle upright and making the hair stand on end. This reaction also happens to animals, like the cat on the left. It has a useful function: When its fur stands up, the cat looks bigger to scare its enemies. The effect also helps animals keep warm, because a fluffed-up coat provides insulation from the cold.

Struwwelpeter

During the 19th century, German writer Heinrich Hoffmann wrote a collection of moral tales to warn children about the dreadful things that could happen if they misbehaved. One of the stories was about a little boy named Struwwelpeter, which means "slovenly Peter" in German. The boy let his fingernails grow for a year and never ever combed his hair. Before long, poor Struwwelpeter ended up looking pretty disgraceful!

Animal Nails

Horse Hoof

Eagle Talons

Goat Horns

A horse's hoofs have to be strong enough to support its weight. An eagle's talons have to be powerful enough to hook and hang on to prey. A goat's horns need to be tough because the animal uses them as defensive weapons. What's the ideal material for building all these features and many others found in animal bodies? The same thing that's in your fingernails: keratin! Keratin comes in many shapes. It forms a bird's feathers and beak, as well as the scales that cover reptile bodies and the horn that forms deer antlers.

Just Nails

Nails grow at an average rate of up to 0.5 cm (0.2 in.) each month. They grow faster in warm weather than they do in cold, and they grow more when you're sick. The record for "longest nail" belongs to Shridhar Chillal of India, whose thumbnail—taped into a spiral in this picture—measured 1.4 m (4 ft. 8 in.)! Chillal last cut his nails in 1952.

The Sense of Touch

Just think what your life would be like if you had no sense of touch. You wouldn't feel pain when something cut or bumped your skin. You wouldn't pull your fingers back when they got too close to a hot stove. You might wander around in the freezing winter air without thinking to put on warmer clothes. Your sense of touch protects you from the dangers of the outside world. Touch also gives pleasure: Think of petting a cat or getting a hug from a loved one.

Special **organs** called **receptors,** found below the skin, decode these sensations of pain, temperature, and pressure. The receptors are part of an intricate network of **nerves** that gather information about the world outside the body and send it to the brain.

Skin Up Close

Receptors sensitive to pain and heat, called free nerve endings, reach close to the surface of the skin. All receptors change physical sensations into electrical signals that are carried to the brain by nerves.

When something cold, like an ice cube, touches the skin, receptors detect it and fire off nerve impulses to the brain. The brain will then tell the body to do something appropriate—such as shivering.

Receptors sensitive to heavy pressure are found deep in the skin. There are lots of them on the palms of the hands and soles of the feet.

Light touch receptors are found just below the **epidermis,** mostly in areas of the body with little or no hair. They can feel the gentle brush of a feather.

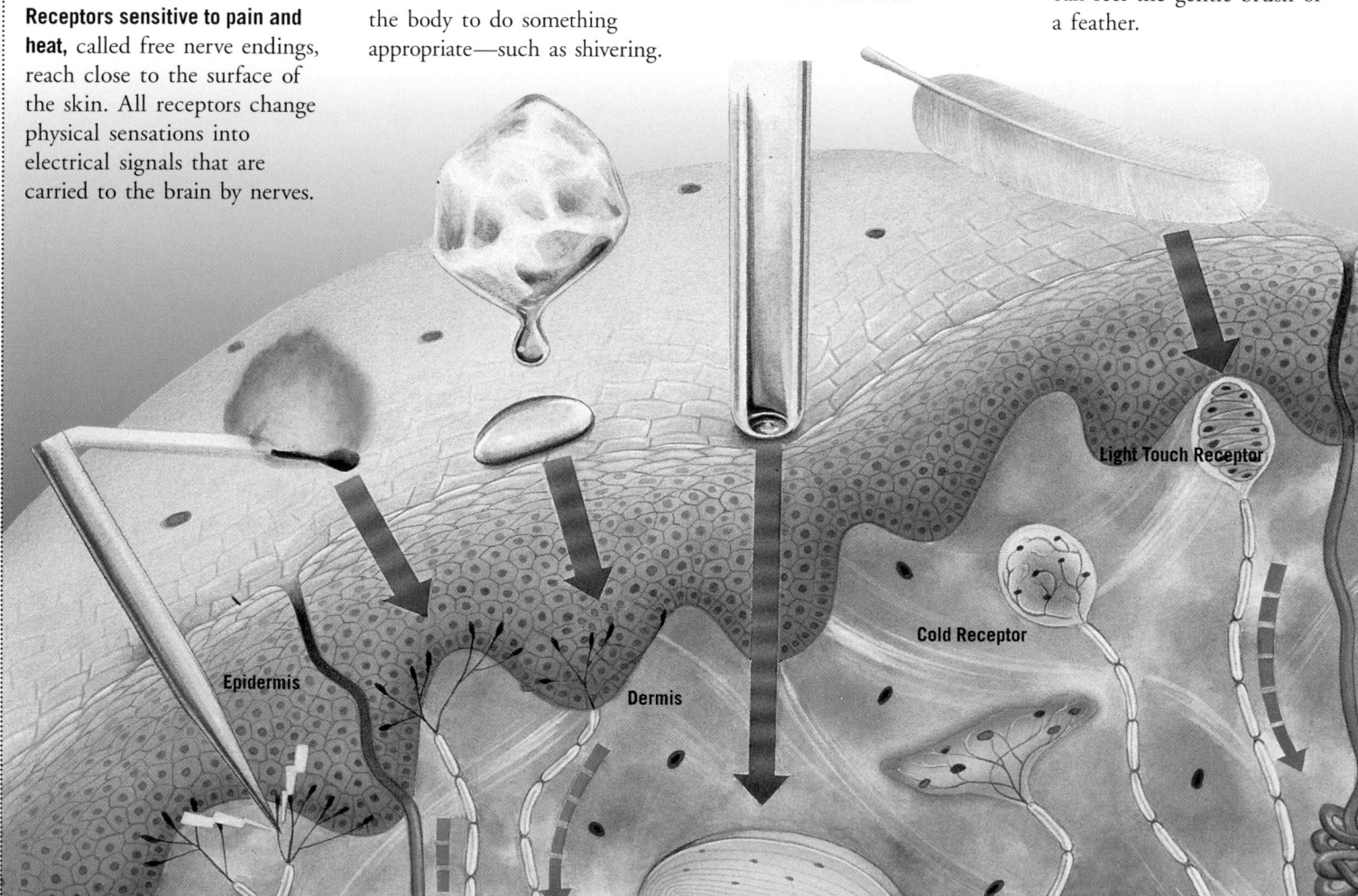

Epidermis

Dermis

Light Touch Receptor

Cold Receptor

Free Nerve Endings

Heavy Pressure Receptor

The Feeling Body

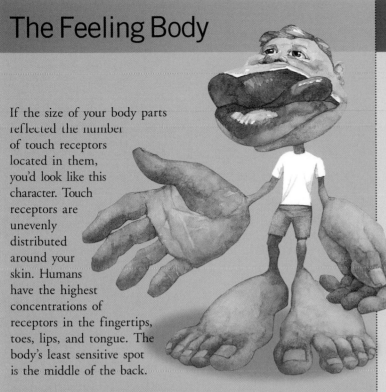

If the size of your body parts reflected the number of touch receptors located in them, you'd look like this character. Touch receptors are unevenly distributed around your skin. Humans have the highest concentrations of receptors in the fingertips, toes, lips, and tongue. The body's least sensitive spot is the middle of the back.

Hot or Cold?

You can trick your sense of touch into sending confused messages to your brain. Try it! Put some water and some ice cubes in one container and stir until the water gets very cold. Put cool water in the second container. Dunk your hand in the icy water and leave it there for about a minute. Take it out quickly and put it in the other container. You may be surprised to discover that now the cool water feels warm to the touch.

Reading Braille

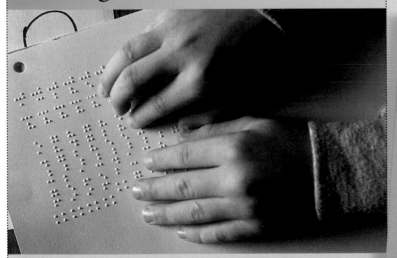

The sense of touch is vital for blind people; it allows them to read Braille. Frenchman Louis Braille, who lost his sight at age three, came up with the system in the 1820s while still a teenager. You read Braille by moving your fingertips—among the most sensitive parts of your body—over raised dots on paper. The dots come in groups of up to six, each group standing for a number, letter, or group of letters. Braille also includes punctuation marks and music notation. Millions of blind people around the world use Braille to read and play music.

Would You Believe?

Ouch Couch!

A yogi, or holy man in India, lies comfortably on a bed of nails. How can he feel no pain? The reason is that the yogi's tough religious training involved learning to control physical sensations, including pain messages sent from the nerves to the brain. Doctors have studied yogis' practices to discover how people can learn to control pain with meditation, not medication.

Eyes Windows on the World

Vision—the ability to see the world—may be the body's dominant sense. The act of seeing occupies about two-thirds of the brain's conscious attention. Two-thirds of the information stored in the brain originally came in through the eyes as pictures, written words, and other images.

The eye is the instrument of vision. It collects information about space, light, color, size, and much more and sends it to the brain for processing. Eyes can't work without light, though. After bouncing off objects all around us, light enters the eye. It passes through a clear outer layer called the cornea to the **pupil** and then to a lens. The lens projects the rays onto the retina at the back of the eyeball. There, special **cells** that are sensitive to color and light convert the image into electrical impulses that travel through the **optic nerve** to the brain. The brain then decodes the messages as images—and sees. All of this complex action happens in a fraction of a second.

Hardworking Muscles

To see clearly, the eyes move up to 100,000 times a day. Six **muscles** work in pairs to control the eyeball. The muscles of each eye are coordinated so that they move together.

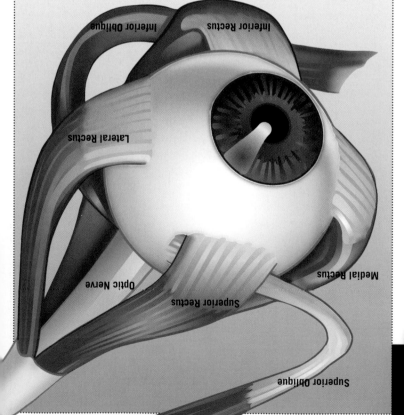

Superior Oblique

Optic Nerve

Superior Rectus

Medial Rectus

Lateral Rectus

Inferior Rectus

Inferior Oblique

A Look Inside

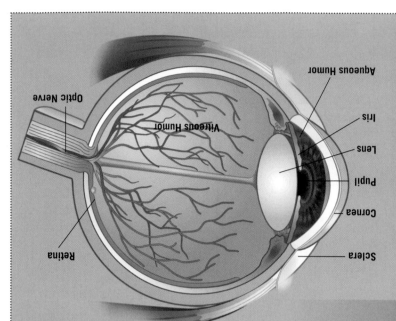

Optic Nerve

Aqueous Humor

Vitreous Humor

Iris

Lens

Pupil

Cornea

Retina

Sclera

Light enters the eye through the clear, domed cornea and moves through the aqueous humor, a clear fluid behind it. Then the light passes into an adjustable hole called the pupil. A ring of muscle around the pupil, called the iris, controls its size. It widens or narrows the pupil depending on how much light enters it. (The iris is also the colored part of the eye.) Finally, the light rays travel

through the lens, which focuses them on a layer of cells at the back of the eyeball called the retina. A human eye is only 2.5 cm (1 in.) wide, yet it contains about 130 million light-sensitive cells in its retina alone.

Inside the eyeball is a jellylike, clear fluid called the vitreous humor. The sclera, the eyeball's tough outer coating, holds in the vitreous humor. The sclera is visible as the white of the eye.

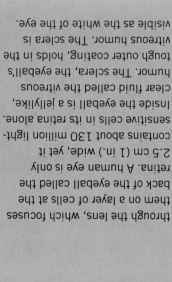

How the Eye Sees

Like the lens in a camera, the lens in your eye captures light from the outside world and focuses it—in this case, onto the retina. Images on the retina arrive upside down because of the way the eye's lens handles light. But the brain automatically flips the images right side up for you.

The retina has two types of light-detecting cells. They are called rods and cones because of their shapes. Each eye contains 120 million rods and six to seven million cones. Rod cells are very sensitive, even in nearly complete darkness, but they can detect only shades of gray. Rods form outlines or silhouettes of objects. Cone cells detect details and color, but they don't work very well in low light. That's why your vision dims and colors seem to fade as the sun goes down in the evening.

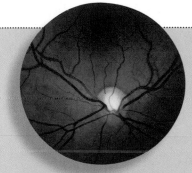

The retina is covered by a complex network of blood vessels *(above)*. The yellow spot marks the location of the optic nerve.

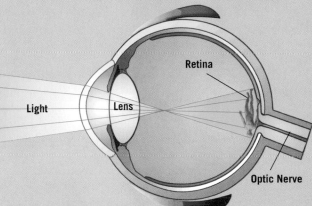

Object
Light
Lens
Retina
Optic Nerve

In a picture taken through a scanning electron microscope *(above)*, you can clearly see some of the cone cells *(blue)* and the more numerous rod cells *(pink)* that line the surface of the retina.

3-D Vision

Because we have two eyes that see objects from slightly different angles, as well as a brain that can put different angles together, we can perceive depth. Each of your eyes sends your brain a different but overlapping image. Your brain then combines both images into a single one that has three dimensions: height, width, and depth.

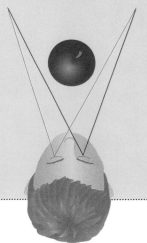

Let's Compare

The pupil is the gateway to the eye. When it's dim out, the pupil expands to let in lots of light for maximum vision. In bright light, the pupil contracts to protect the sensitive cells inside your eye.

Dim Light—Pupil wide open

Bright Light—Pupil small

The giant squid, one of the largest creatures in the world, can grow to be more than 18 m (60 ft.) long. Its huge eyes, the largest of any animal's, are the size of a kid's head. It needs big eyes to see in the dark, murky waters 600 m (2,000 ft.) below the ocean surface, where it lives.

How Big?

Ol' Blue Eyes

The Sense of Sight

We use our sense of sight for both information and delight. Sight creates images of our surroundings. Those images give us information we use to make decisions, such as how far to step when we're climbing stairs. We use sight to learn as we read books. We use it for enjoyment when we look at a painting or admire a sunset.

Sight allows us to recognize the people we love and to find out about people we don't know. Sight lets us "read" others' faces to see how they're feeling. Because sight involves seeing distances and where things are located in space, it helps us dance, exercise, or drive from place to place. Sight helps us experience the world.

Sight is different from perception, however. Perception means making sense of what you see. Sometimes your perceptions can fool you, as you'll see below.

Color Blindness

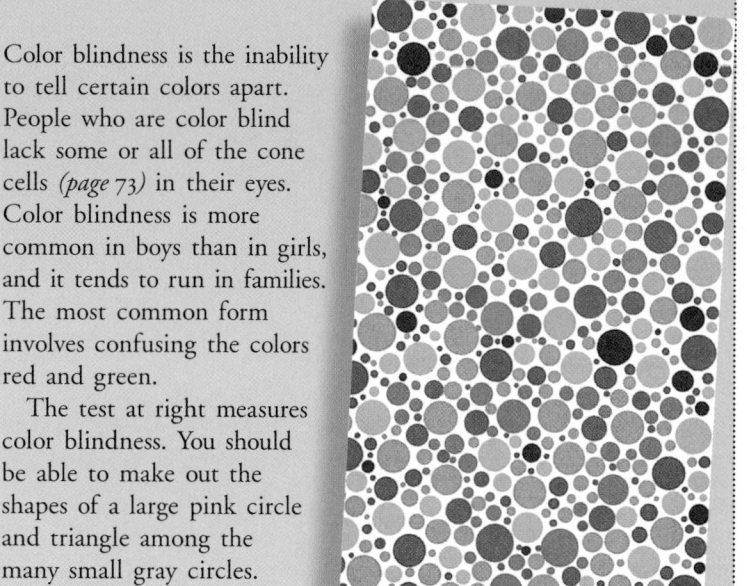

Color blindness is the inability to tell certain colors apart. People who are color blind lack some or all of the cone cells *(page 73)* in their eyes. Color blindness is more common in boys than in girls, and it tends to run in families. The most common form involves confusing the colors red and green.

The test at right measures color blindness. You should be able to make out the shapes of a large pink circle and triangle among the many small gray circles.

Eye Fooled You!

Tall—or small? As this photo shows, our idea of size can be a matter of perspective. The room in the picture is distorted. In reality, the back wall recedes to the left, but the windows and doors are placed to make us think that the room is rectangular. We're so used to rectangular rooms that we expect the two figures to be on the same plane. In fact, the child on the left is much farther back, and therefore looks smaller. But our perceptions trick us, making one girl look like a giant and the other tiny. In reality, both kids are the same height.

When Vision Is Blurry

Are you nearsighted? Farsighted? Blame it on your eyeball. The shape of your eyeball determines how clearly you see. Normally, the lens at the front of your eye focuses images on the retina at the back of your eye. If the eyeball is too long, images focus in front of the retina, making faraway things look blurry. If the eyeball is too short, images focus behind the retina, making close things blurry. Both problems can be corrected with glasses or contact lenses.

Nearsighted

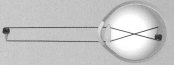

Uncorrected

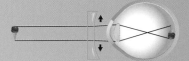

Corrected

Farsighted

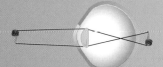

Uncorrected

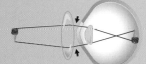

Corrected

In 1880 a healthy baby girl was born in a small Alabama town. When she was less than two years old, Helen Keller fell ill and became deaf and blind. After searching for ways to communicate with Helen, her parents found teacher Anne Sullivan *(below, right)*. Anne Sullivan taught the seven-year-old by spelling words with her fingers in Helen's hand. For example, she spelled d-o-l-l, then gave Helen a doll. Helen began to understand that the words were names that stood for objects. Later, at school in Boston, she mastered Braille and learned to speak. At college she wrote *The Story of My Life,* which still inspires people today.

The Eyes Have It

Eagle Eye

Did you know that the golden eagle has the sharpest eyesight in the world? The bird can spot rabbits and other prey on the ground far below from a distance of more than 0.80 km (0.50 mi.). Experts think that golden eagle sight is seven times as acute as human sight! No wonder mice run for cover when they sense an eagle hovering overhead.

The Third Eye

In Hindu belief, the god Shiva *(right)* has not two but three eyes. The third eye, on his forehead, shows that he has a spiritual sight, or "second sight," as well as everyday sight. Many cultures have nurtured the idea of an invisible third eye with mysterious powers, including clairvoyance, the ability to predict the future.

Ears Shaped for Sound

Our ears are sound collectors, gathering sound waves from the air. They allow us to hear everything from whispers to booms. Our ears are also sound interpreters, translating sound waves into electrical signals to send to the brain. They work with the brain to create the sense of hearing. Our ears are even balancers, helping us to keep our body in position.

Hearing begins when sound waves gathered by the curve of the outer ear move into the head through the ear canal toward the middle ear. In the middle ear, a membrane called the eardrum vibrates, passing sound waves on to the inner ear. The inner ear contains a tiny snail-shaped structure called the cochlea that is filled with fluid. Inside the cochlea, information from sound waves changes into electrical signals for the brain. The brain gives meaning to these sounds—"whisper," "crash," "music," or "crying baby." Attached to the cochlea are three C-shaped structures, the semicircular canals, which help the body keep its balance.

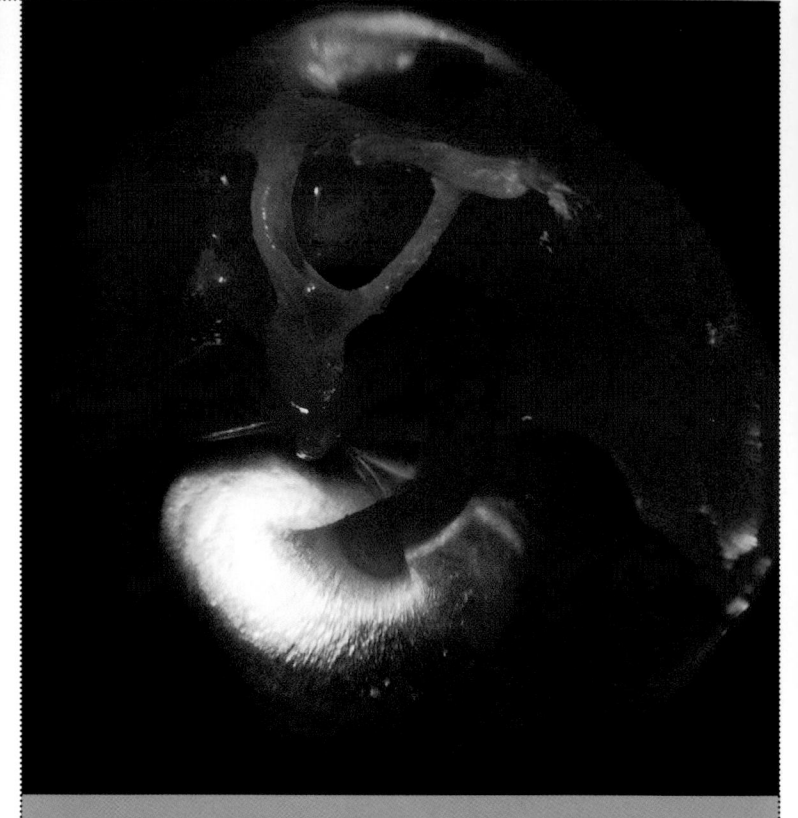

Looking from the inside out at eight times normal size, you can see the tiny bones of the middle ear. The hammer is attached to the inside of the eardrum, lighted up for this photo. It strikes the anvil, which in turn sets the stirrup bone vibrating. This sequence passes sound waves along to the inner ear. The stirrup, seen in the foreground at the top of the picture, is the smallest bone in the body.

Inside the Ear

When you think of an ear, you probably picture the pinna, which is the scientific name for that flap of skin and **cartilage** on the outside of your head. But much of the intricate ear system is hidden deep inside. The most sensitive parts are protected by the skull's thick scaffolding of bone. These are the delicate anvil, stirrup, and hammer bones and the spiral-shaped cochlea, which translates sound waves into **nerve** signals.

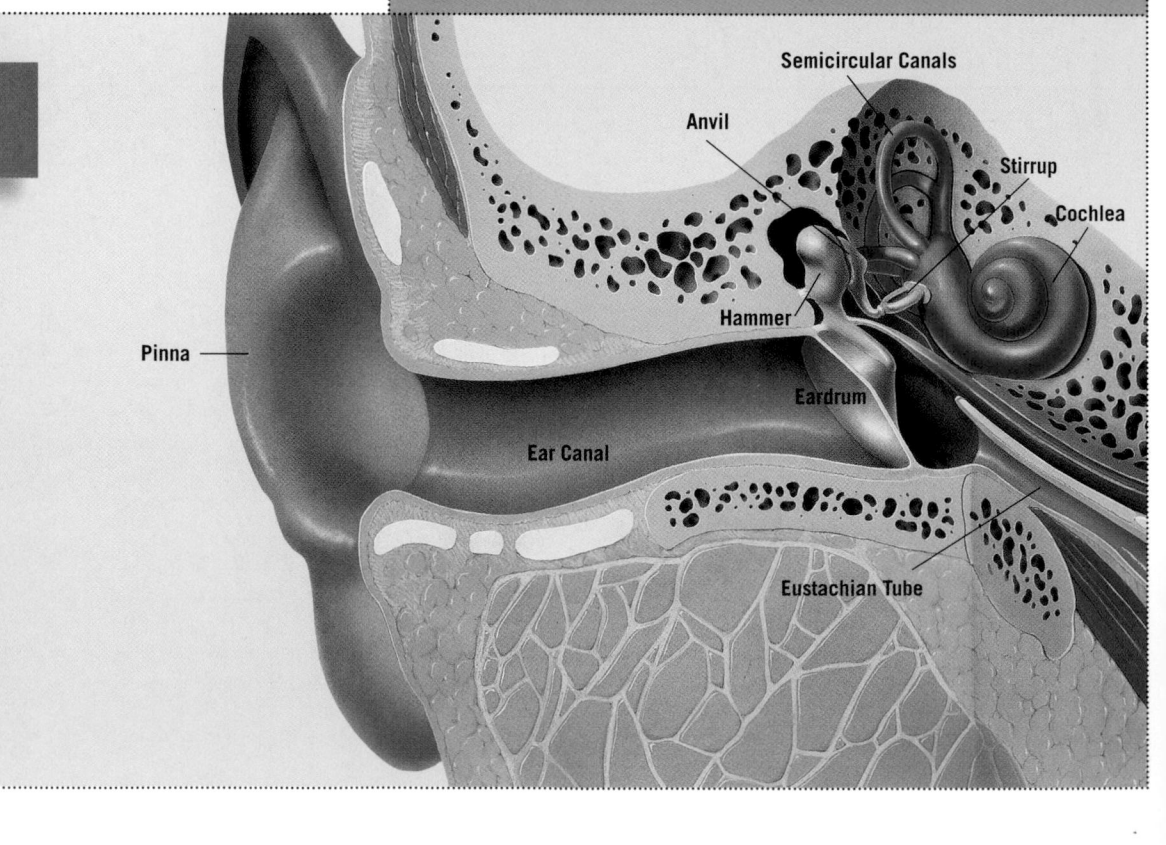

Semicircular Canals

Anvil

Stirrup

Cochlea

Hammer

Pinna

Eardrum

Ear Canal

Eustachian Tube

Balancing Act

Deep within the ear are three tubes called the semicircular canals *(diagram, page 76)*. These tubes, positioned almost at right angles to one another, are filled with a jellylike fluid. When you move your head and body—whether you're rushing off to school or balancing upside down like this acrobat—the fluid moves, too. As it sloshes to and fro and up and down, it touches patches of nerves that let your brain know where you are in space. Your brain makes sure your **muscles** respond to keep you from falling.

Why Ears Pop

Normally, the pressure of air pushing on the eardrum is the same inside and outside the ear. But when you go up in a plane the pressure inside the cabin drops. The pressure inside the ear is now greater and it causes the eardrum to bulge out *(top right)*. Your ears hurt. If you swallow, your Eustachian tube opens up and releases some of the pressure *(bottom right)*. You hear a "pop" as your eardrum snaps back to its normal position.

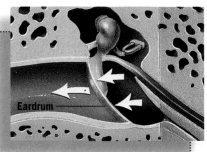

Eardrum

How We Hear

Sound waves constantly travel from the outside world into the ears. After the waves vibrate the eardrum and are moved along by the auditory bones, they reach another stretched **membrane** called the oval window. This membrane transmits the vibrations through the fluid-filled cochlea *(shown uncoiled below)*. Inside the cochlea are thousands of sensory hairs arranged in rows. Each hair **cell** has about a hundred bristles that translate the movement of the fluid passing over them into messages to the brain.

Let's **Compare**

Tiny hairs line the inside of the cochlea. These delicate hairs are responsible for changing sound waves into signals that the brain can interpret. If the hairs are damaged or missing, a person will be partially or totally deaf. Illnesses or infections can damage hairs in the cochlea. Old age weakens them. The hairs can also be badly hurt by very loud sounds.

Deaf Ear

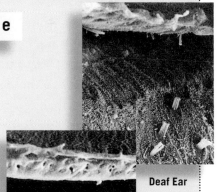

Healthy Ear

Listen Up!

The ears of the fennec fox, which can measure up to 15 cm (6 in.) long, act like satellite dishes. When it hunts, the little animal swivels its ears to pick up the tiniest sounds made by its prey, even if they're coming from under the desert sand.

Auditory Bones

Oval Window

Cochlea

Eardrum

Sound Waves

The Sense of Hearing

A siren wails. Your best friend whispers. A guitar screams out a high E flat. Like the other senses, your hearing brings you warnings, information, and pleasure. Hearing helps you learn. It allows you to make decisions, stay safe, and have fun. You use your hearing when you listen to your friends and family, your teachers and coaches. Hearing lets you listen to a symphony, as well as to nature's music, such as singing birds and pounding waves.

Hearing also acts as an early-warning system. When you hear thunder, you know a storm is coming, so you look for shelter. When you hear an automobile horn or a police car siren, you know it's time to get out of the way. Hearing a fire alarm tells you to find a safe escape route. Your hearing is so important that it's vital to protect it.

How Loud Is Sound?

Scientists measure sounds in units called decibels. Decibel levels rise fast; every increase of three decibels means a doubling of loudness. And loudness can permanently hurt your ears. Sounds of 120 decibels or more are dangerous to your hearing. They can cause permanent hearing loss.

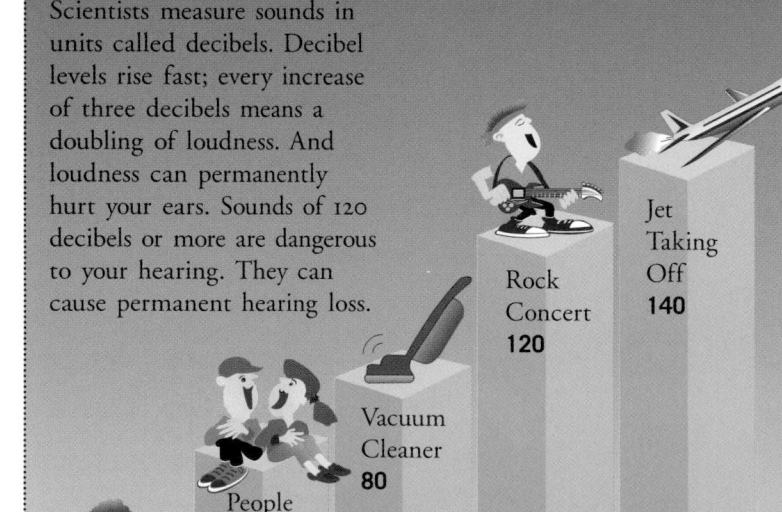

Whisper
20

People Talking
60

Vacuum Cleaner
80

Rock Concert
120

Jet Taking Off
140

Earsplitting Sound

Did you know that a typical rock concert is hazardous to your health? Amplified sound blasting from giant speakers is loud enough to harm your hearing.

The 120 decibels that pour from the stage are in the range that can cause serious damage to sensory hairs in your inner ear. If you go to a rock concert, wear earplugs. That way, you'll be able to go on enjoying music in the future!

If you use a personal radio or CD player, turn the volume down. By directing sound into your ear, headphones can produce levels loud enough to hurt your ears. Hold your headset an arm's length away. If you can hear the sound, the volume is too high.

Noise pollution is also a growing problem in the modern world. For example, a jet aircraft taking off roars at 140 decibels. Many cities now have rules to keep down noise.

Hearing Range

If you have a pet, you know it hears sounds that you miss. Animals, including humans, hear sounds only within a certain range of sound waves. Scientists use a unit called the hertz (Hz) to measure these waves. The greater the number of hertz, the higher the sound. Humans hear sounds in the 20- to 20,000-hertz range. Many animals—including bats and dogs—detect sounds whose pitch is far higher than those we hear.

Vibrations per Second (Hz)

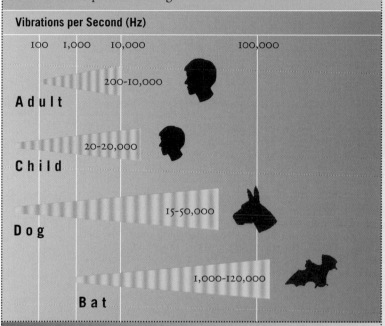

| 100 | 1,000 | 10,000 | 100,000 |

Adult 200-10,000

Child 20-20,000

Dog 15-50,000

Bat 1,000-120,000

Hearing Dogs

You may know about Seeing Eye dogs, which guide the blind. Did you know that dogs also help deaf people? The intelligence and loyalty that make dogs good guides for the blind also aid hearing-impaired people.

Dogs are trained to alert the humans they work with to everyday sounds such as door-bells, alarm clocks, telephones, and fussing babies. The canine companions also inform deaf people of sounds that mean danger, such as smoke alarms or sirens.

At least one such dog is a real hero. Mr. Bounce *(left)*, a Pomeranian, saved his owner's life twice—once when a carbon monoxide detector went off, and again when he scared off a burglar.

People Ludwig van Beethoven

About 1800, when the great German composer Ludwig van Beethoven was in his late 20s, he noticed a humming in his ears. Gradually he realized that he was growing deaf. Modern doctors believe that Beethoven had a condition called otosclerosis. Today he could have been cured, but not back then. Beethoven could not hear his later music, but he could imagine the notes in his head. He also sawed off the legs of his piano, playing it lying down so he could feel the vibrations through the floor. While deaf, he created some of his greatest works, including the famous Ninth Symphony. When he conducted the first performance of this symphony, he could not hear the loud applause. A member of the orchestra made him turn around to face the audience. Only then did he see everyone clapping.

Then & NOW!

Devices to amplify sound, or make it louder, have changed a great deal. In the 1600s, people made horn-shaped aids to gather sound and funnel it into the ear. By the beginning of the 20th century, the development of electric hearing aids had begun. Recently, scientists have made powerful transistor hearing aids that are small enough to fit right into the opening of the ear. These little devices pick up sound from the air, amplify it, and send it directly into the ear.

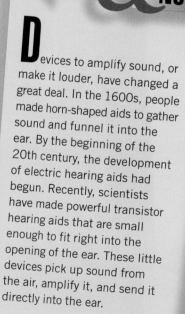

The Sense of Smell

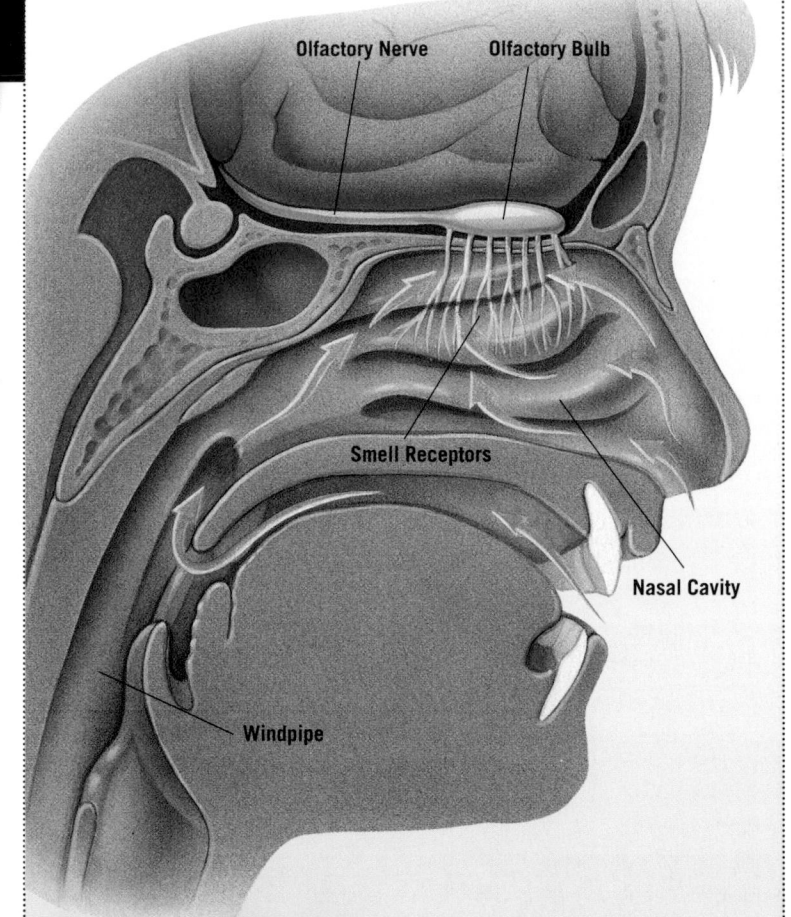

Olfactory Nerve   Olfactory Bulb

Smell Receptors

Nasal Cavity

Windpipe

Mmmm . . . chocolate chip cookies!" you say as you walk into the kitchen. How do you know what's cooking? Your nose knows! Scientists call smell a chemical sense, because the nose detects **molecules** of certain chemicals floating in the air. When the molecules land inside the nose, they trigger special patches of **cells** there. The cells send **nerve** impulses to the olfactory bulbs, which sort them out and send them to the brain via the olfactory nerves.

For humans, as for other animals, smell is a way of checking out the environment. Pleasant-smelling food appeals to us and makes us hungry. Rotten food and drink, which could make us sick, repel us. Smell also warns us of dangers, such as the smell of smoke from a fire or air filled with exhaust fumes.

Sometimes smell can recall memories and emotions. The part of the brain that handles smell is closely connected to the part that handles mood and memory.

Sniffing Helps Smelling

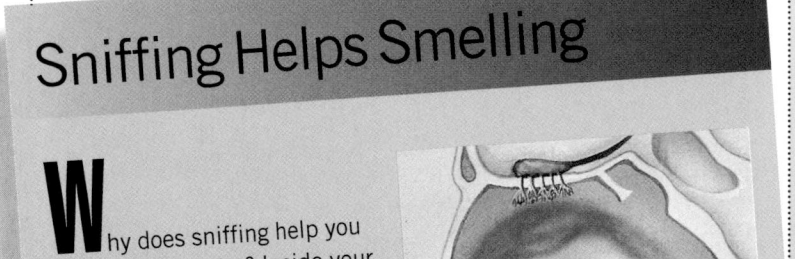

Why does sniffing help you smell things better? Inside your nose are two patches of fine hairs no larger than postage stamps. Each patch holds millions of specialized cells for detecting scent molecules. Nerves connected to these cells then send odor information to the brain. Unlike normal breathing (*bottom right*), a good sniff (*top right*) pulls air to the top of the nasal cavity—bringing more odor molecules up to the smelling cells.

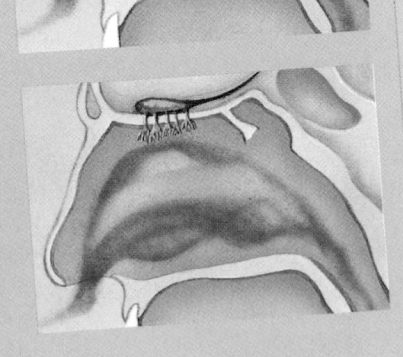

How the Nose Smells

Your nose has two important jobs. First, it is a main gateway for air into your **respiratory system** (*pages 42-43*). Air enters through the nostrils and gets warmed up and cleaned up in the nasal cavity before being sucked down the windpipe into the lungs. Air can also flow up into the nasal cavity from the mouth.

The nose's second job is smelling. Supersensitive smell **receptors** located in the roof of the nasal cavity have tiny hairs that hang down and pick up odors as they waft by. They send "smell" messages to the olfactory bulbs, which send them on to the brain. An ordinary person can distinguish about 4,000 different odors!

The Nose Knows

Smell specialists make their living detecting and analyzing scents. These experts train themselves to recognize a wider range of smells than most people can—as many as 10,000 odors! They test everything from perfumes and cosmetics to the effectiveness of underarm deodorants. Wine-makers, chefs, and many other professionals need sharp smelling skills.

Would You Believe?

In the Middle Ages, physicians believed that diseases could be spread by foul smells. This doctor is dressed from head to toe in leather to shield him from the smell of victims of the Black Death, or plague. His beaked mask holds perfume and spices to mask the odor of disease.

Bloodhounds

Bloodhounds have no rivals when it comes to following scents. The bloodhound's sense of smell is hundreds of times more sensitive than a human's. Bred centuries ago as hunting dogs, bloodhounds have found lost children and helped police departments track down criminals.

The dog's wide nostrils point forward and down, allowing it to pick up scents on air currents rising from the ground. Its nose has 220 million scent receptors! All the bloodhound has to do is sniff something belonging to the person it's looking for. Then it's off on an invisible track that no one else can see—or smell.

"In her own Words"

Have you ever whiffed something and remembered an event or person from long ago? Scientists have proved that scent and memory are closely linked in the brain. Helen Keller (page 75), who was both blind and deaf, knew all about the emotional power of smell.

"Smell is a potent wizard that transports us across a thousand miles and all the years we have lived. The odor of fruits wafts me to my Southern home, to my childish frolics in the peach orchard. Other odors, instantaneous and fleeting, cause my heart to dilate joyously or contract with remembered grief. Even as I think of smells, my nose is full of scents that start awake sweet memories of summers gone and ripening grain fields far away."
—Helen Keller, *The World I Live In*

The Sense of Taste

T aste is more complicated than you might think. When you taste something, you're actually sensing four qualities that make up a food's flavor: its smell, temperature, texture, and taste itself, as measured by your taste buds.

When you eat or drink, chemicals in the food dissolve in your **saliva.** Special receptor cells on your tongue and around your mouth—your taste buds—detect the chemicals. Different buds respond to different tastes, such as sweet or salty, in the chemicals. The sense of smell also comes into play. The aroma of the food rises into your nose, which sends messages to the brain; the smell of your food is a large part of its flavor. (That's why, when you have a cold and can't smell things well, your food doesn't taste very good.) Taste preferences vary enormously, from culture to culture and person to person. Some people love to eat fiery curries; others prefer bland food. What is your favorite taste?

Papillae

T he little lumps and bumps that cover your tongue are not taste buds—they are called papillae. The rough surface they create helps the tongue grip food. Fungiform papillae (the larger dots in this picture) are shaped like mushrooms. Filiform papillae are frilly and hairlike.

Taste buds are tiny groups of cells, much smaller than papillae. About 10,000 of them are scattered around the tongue. They are found in the sides or top of some papillae.

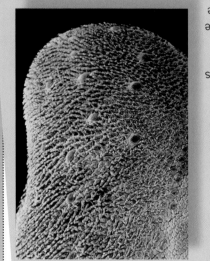

A Tasty Tongue

Your tongue is a strong **muscle.** It flexes in many ways to help you mash up and move food from your mouth to your throat. It even helps you form the words to ask for a second helping! The tongue is the major **organ** of taste, thanks to microscopic, onion-shaped bunches of cells—the taste buds—that take care of that job. Taste buds detect flavor and send signals to the taste centers in your brain.

Your taste buds sense four basic flavor groups: sweet, salty, sour, and bitter. All other tastes are combinations of these four. The back of the tongue has the largest concentration of taste buds for bitter tastes. The sides of the tongue detect sour flavors. Buds that recognize salty tastes are found mostly on each side of the tip. At the tip of the tongue are the buds that sense sweet foods.

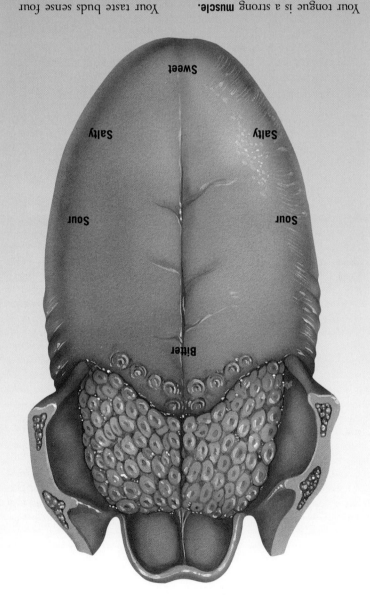

Sweet

Salty

Salty

Sour

Sour

Bitter

Test Your Taste Buds

Try it!

Test your taste! With the help of a parent, find food with four distinct flavors: sweet, salty, sour, and bitter. Below you see a cupcake for sweet, potato chips for salty, sliced lemon for sour, and a cup of tea for bitter. Place each kind of food on four places on your tongue: the tip, the back, the sides in back, and the sides in front. Which food tastes the strongest in which location? Are there places on your tongue where you can't taste certain flavors?

Scary Tongue Talk

In New Zealand, sticking out your tongue is not just rude—it can be a declaration of war. The native Maori people of New Zealand used to perform a war dance and chant called a *haka* before going to war. The dance included rolling the eyes and sticking out the tongue to make the warrior as frightening as possible. Today the dance is only done on ceremonial occasions.

Would You Believe?

Sweet Feet

Butterflies and moths taste with their feet. Special cells on the tips of their feet can taste leaves and flowers. A touch-down of just a few seconds is enough to identify if a plant is good to eat or useful as a site for laying eggs.

Strange But TRUE!

Delicious Colors

Can you imagine seeing sounds? Feeling flavors? Tasting colors of words? People with a rare brain condition called synesthesia experience the world this way. Their brain receives mixed-up sense messages, as if the switchboard delivering messages got its wires crossed. A synesthete, as such a person is called, may perceive music as colors, with each note creating a different hue. Others add the sensation of taste as well as color to sound. One synesthete described her husband's laugh this way: "It's a wonderful golden brown, with a flavor of crisp, buttery toast, which sounds very odd I know, but it is very real."

Sweet Tooth

Sweet treats and small kids seem to go together naturally. There's a biological reason for that: Children younger than five have many more taste buds than adults do—and they're on the lips as well as the tongue. Kids seem to be more sensitive to strong tastes, and more inclined to like sweets.

What Is Digestion?

Like a car, the body needs fuel to keep going. The job of the **digestive system** is to convert the food you eat into the energy, or fuel, that your body's **cells** need to perform their jobs. This fuel also provides **nutrients** that the body needs to grow, maintain, and repair **tissue.**

Digestion begins as soon as you take a bite of food and chew it up. Once you swallow, the process continues in the stomach, which mixes up your meal with powerful acids and **enzymes.** These chemicals break the food down into material that the body can absorb. Most absorption takes place in the small intestine, where nutrients from food pass into the bloodstream to be delivered to the cells. Then what's left of the food passes into the large intestine, whose main job is to remove water from the food by absorbing it for the body to use. Parts of food that can't be used by the body pass into the rectum and are expelled as waste when you go to the bathroom.

Famous 1 FIRSTS

Watch What You Eat!

In 1822, U.S. Army surgeon William Beaumont was asked to treat a patient, Alexis St. Martin, who had been accidentally wounded by a shotgun blast. St. Martin survived, but the 6.4-cm (2.5-in.) hole blown in his stomach would not close completely. Through this amazing peephole, Dr. Beaumont studied the digestive process in action by peering into St. Martin's stomach through a tube. Once he even inserted an oyster tied to a string; when he pulled the string out an hour and a half later, the oyster was gone!

Teeth
Tongue
Esophagus
Liver
Stomach
Pancreas
Gallbladder
Large Intestine
Small Intestine
Rectum

The Digestive Tract

The digestive tract is a kind of complicated tube that, if all stretched out, would be about 9 m (30 ft.) long. The food you eat makes several stops at places in this tube. Processes that happen at these different locations help to move the food, break it down, and absorb nutrients to fuel the activity of your cells. During the journey, food is changed from a solid—such as slices of apple—into a mushy substance that resembles a thick milk shake. The body's cells absorb and use chemicals from this material. Some foods take longer to digest than others, but in general the whole process takes between 19 and 36 hours to complete.

Adding Juices

As food leaves the stomach and enters the top of the small intestine it is flooded with digestive juices from the pancreas and the gallbladder.

Top of the Tract

In just seconds, the teeth grind up a bite of apple. The tongue helps mix it with **saliva** pumped into the mouth from the salivary glands.

Down the Hatch

It takes five to eight seconds for **muscle** contractions in the throat to move the chewed food down the **esophagus.**

Into the Mixer

Muscles in the stomach churn food with strong acids for four to six hours to turn it into a thick, soupy substance called chyme.

Last Stop

For 12 to 24 hours, the large intestine stores the food material and soaks up all remaining nutrients and water from it. What is left over is then expelled from the body.

Small Name, Big Job

The small intestine is a long, coiled tube lined with villi, tiny bumps that absorb nutrients over three to six hours and send them into the bloodstream.

Heavy Metal Diet

Would **YOU** *Believe?*

You may have heard people say they're hungry enough to eat a horse. But a two-seater airplane? That's exactly what 28-year-old Michel Lotito of France did in 1978. He chopped a Cessna into fingernail-size bits, then ate them one by one. Lotito, who can consume up to 1 kg (2 lb.) of metal a day, travels around the world eating such objects as bicycle spokes and shopping carts. In recognition of this remarkable talent, *The Guinness Book of World Records* presented him with a brass plaque. He ate it.

A Look in the Mouth

Y ou might not think of your mouth as part of the digestive tract, but that is where the whole process of digestion begins. Each part of the mouth has a specific job to do.

Your lips open up to let food in, and close shut to keep it there. The teeth are designed to slice, chop, and grind food into smaller pieces that will fit into the **esophagus.** Since extreme temperatures can damage the delicate lining of the esophagus, your mouth helps to bring the food you eat to body temperature so that it's not too hot or too cold when you swallow it.

As you chew, your muscular tongue moves food around to mix it thoroughly with **saliva.** Saliva lubricates the food to make it easier to swallow, and it contains **enzymes** that begin breaking the food down. Saliva also dissolves chemicals in the food so you can taste it. Finally, the tongue rolls the chewed-up food into a wad called a bolus, which it then moves to the back of your mouth to be swallowed.

Hard Palate

Soft Palate

Uvula

Tongue

Then & NOW!

A nthropologists—scientists who study the history of the human race and how we lived—have found fossil remains of teeth from one of our earliest ancestors, the humanlike *Australopithecus,* who lived more than two million years ago. The remains *(top right)* show that our ancestors had a bigger jaw-bone and much larger teeth than we do. They needed bigger teeth because the food they ate was much harder to chew. It wasn't processed like most of the food we eat today.

Jawbone of *Australopithecus*

Jawbone of Modern Human

Down in the Mouth

Your teeth have different names and are shaped to do specific jobs. The front teeth are called incisors, and these eight wide, flat teeth—four on top, four on the bottom—are designed to bite and cut. Next to them are the pointy canine teeth—four in all—which tear and rip at food. Behind the canines, the flat, broad premolars and molars—five on each side, up and down—grind up the food. Some people have four fewer molars because they don't have the ones known as wisdom teeth.

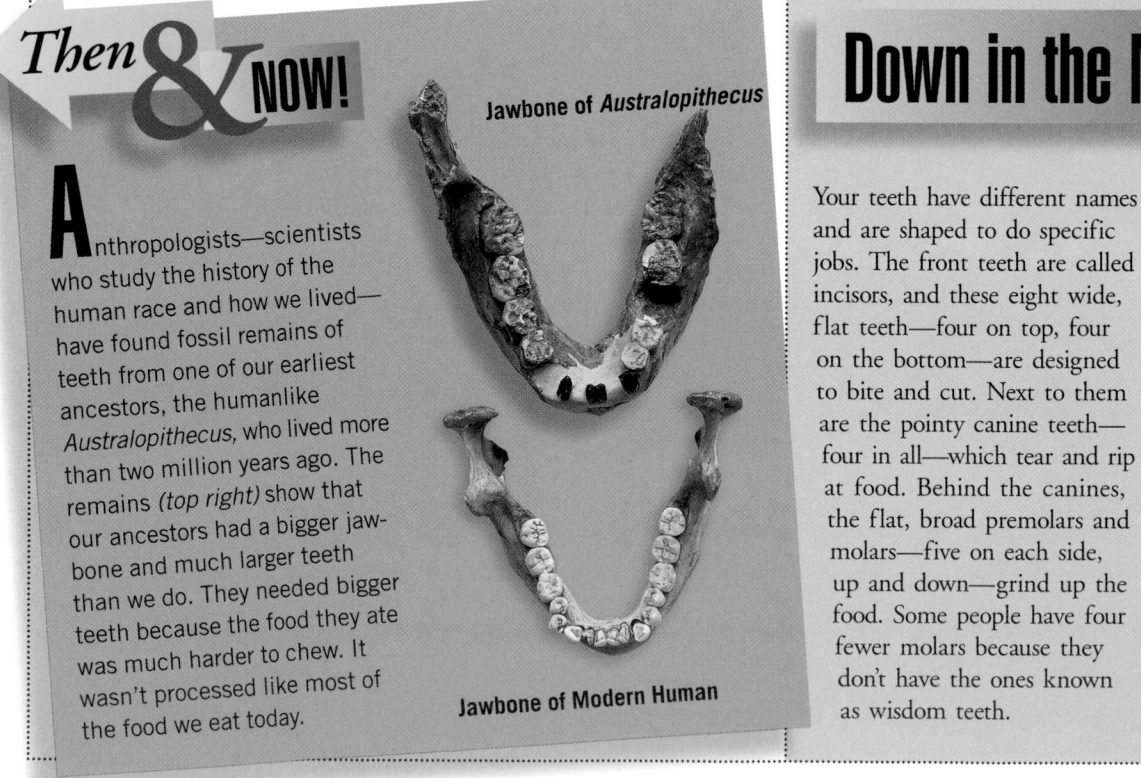

UPPER TEETH

Incisors
Canines
Premolars
Molars

LOWER TEETH

Molars
Premolars
Canines
Incisors

Brush Well!

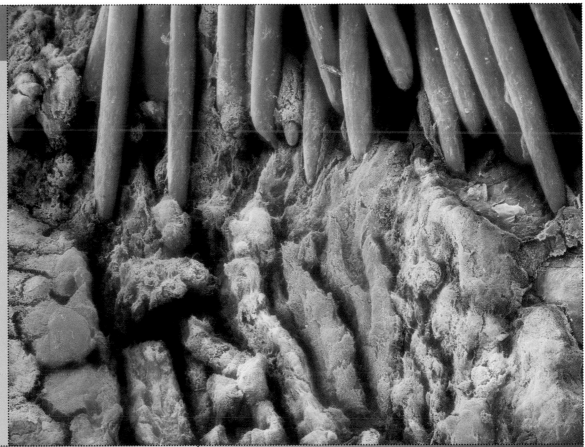

Although a tooth's surface appears smooth, when you look at a tooth through a microscope *(right)* you can see that the surface is actually fairly rough terrain. Food particles can easily lodge in tiny crevices there. **Bacteria** and **mucus** combine with this leftover food to create a sticky coating on your teeth. The bristles of a toothbrush, magnified 75 times in this picture, are designed to loosen and scrub away the coating, called plaque. But if the plaque is allowed to build up, acids that are produced by the bacteria will weaken tooth enamel and cause pockets of decay. These pockets are called cavities.

The Whole Tooth

The shiny white enamel that covers every tooth is the hardest substance in the body—harder even than bone. Designed to protect a sensitive core of **tissue, nerves,** and blood vessels, enamel is made up mostly of **calcium** and phosphorus. Just below it is a hard layer called dentin, which is similar to bone and is nourished by blood vessels. These vessels, along with nerves, are located in the pulp at the center of the tooth. Roots extend from each tooth into the jawbone, where they are held in place by a tough layer called cement.

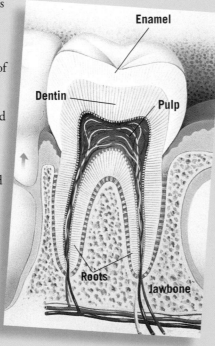

Enamel

Dentin

Pulp

Roots

Jawbone

Why Does My Mouth Water?

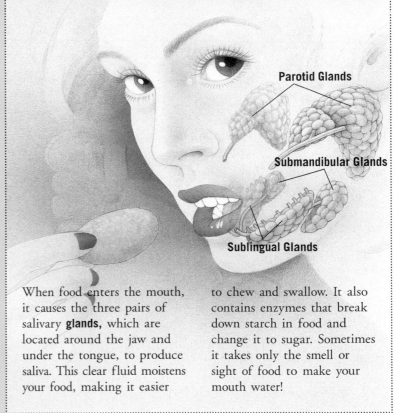

Parotid Glands

Submandibular Glands

Sublingual Glands

When food enters the mouth, it causes the three pairs of salivary **glands,** which are located around the jaw and under the tongue, to produce saliva. This clear fluid moistens your food, making it easier to chew and swallow. It also contains enzymes that break down starch in food and change it to sugar. Sometimes it takes only the smell or sight of food to make your mouth water!

Into the Stomach

After you swallow, your food heads for the stomach—a sturdy bag crisscrossed with three layers of strong **muscles** that expand and contract to hold the food you eat, churn it up, and move it along. An adult stomach measures about 25 cm (10 in.) long and can expand to hold up to 4 l (4.2 qt.) of food and drink when it is very full.

To process food, the stomach lining secretes gastric juices. These contain water, **enzymes,** and other materials, and their job is to break down the food and kill any **bacteria** in it. The most powerful of the gastric juices is hydrochloric acid—a solvent so strong that factories use it to dissolve metal! Fortunately, a protective layer of **mucus** guards the stomach lining from this acid. But, the stomach still loses and has to replace half a million cells a minute! The entire stomach lining renews itself every three days.

By the time food is ready to leave the stomach it has been turned into a thick milk shake-like substance called chyme. The stomach then squirts the chyme into the small intestine a little at a time. It takes between four and six hours to empty the stomach.

The stomach is a J-shaped bag wrapped in bands of muscle that run along its length and also wrap around it. These muscles contract and relax about three times a minute, churning the food with gastric juices produced by the stomach. When the food is thoroughly mixed, the pyloric sphincter—a ring of muscle at the bottom of the stomach—opens and squirts the contents into the small intestine.

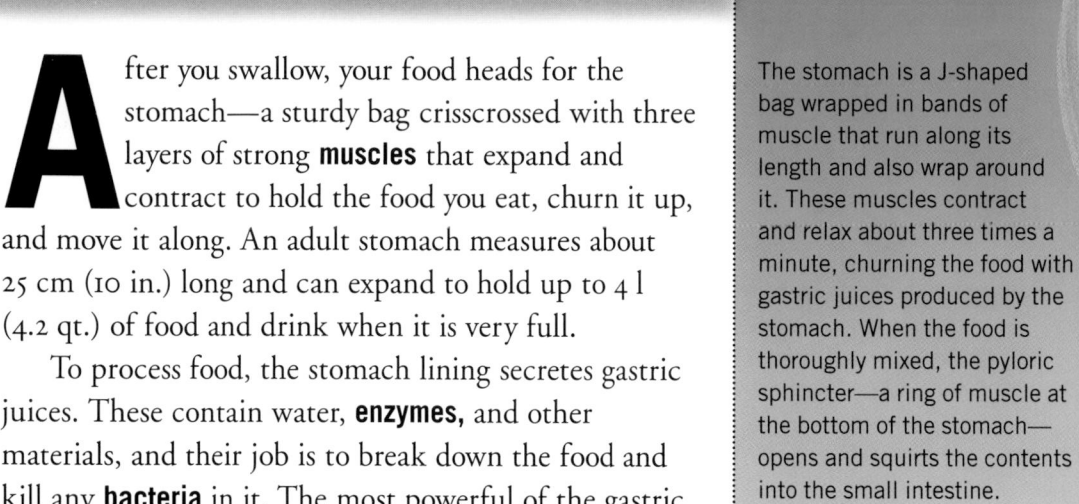

Esophagus

Stomach

Longitudinal Muscle

Circular Muscle

Pyloric Sphincter

Oblique Muscle

Small Intestine

The Food Journey Begins

Did you know that a person swallows 3,000 times a day? Every time it happens, your tongue squeezes the food or drink up and toward the back of your mouth. As you swallow, the nose and the trachea—your airway—are closed off automatically to keep food from entering your nose and lungs.

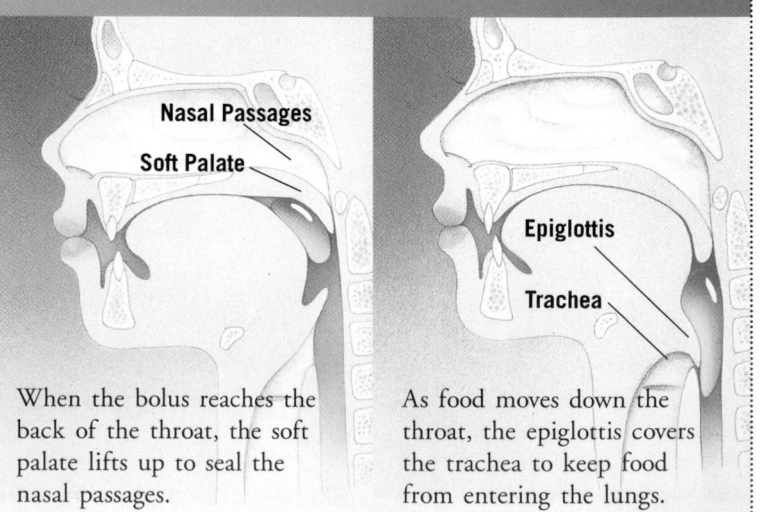

Food Bolus

Tongue

Esophagus

The tongue pushes a bolus, or wad of chewed food, against the roof of the mouth and toward the throat.

Nasal Passages

Soft Palate

When the bolus reaches the back of the throat, the soft palate lifts up to seal the nasal passages.

Epiglottis

Trachea

As food moves down the throat, the epiglottis covers the trachea to keep food from entering the lungs.

Can We Swallow Upside Down?

The **esophagus** is a hollow tube, but when you swallow, food doesn't just fall through it and land in the stomach. It has to be pushed. To do this, the walls of the esophagus and the muscles that line them expand and contract in waves to force the food down. When it reaches the bottom of the esophagus, a valve opens to let the food into the

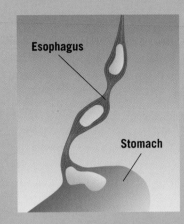

Esophagus

Stomach

stomach. This process, called peristalsis, is forceful enough to move food whether you're standing upright or hanging upside down. But please don't try to swallow upside down; it could make you choke.

The Stomach Lining

The stomach lining lies in folds called rugae *(below)* that expand and contract as the stomach fills and empties. These folds contain the **glands** that produce the gastric juices that break down food. But what keeps the stomach from digesting itself? A thick, sticky mucus covers the inside of the stomach. If not for that coating, the gastric juices would digest the stomach along with the food!

Would **YOU** *Believe?*

A *Really* Full Stomach!

It can take up to 36 hours for a human being to completely digest a meal and distribute its **nutrients** through the body. But a python's digestion takes a lot longer than that. A meal of a large wild pig *(right)* can take a python weeks or even months to digest. No wonder this snake eats only one large prey at a time! After such a meal, many snakes look for sunny, sheltered spots, where they lie still and allow the heat of the sun to speed up the digestive process—and shrink the snake's body back to its normal size!

Inside the Intestines

Food Processors

In the space below the stomach, the body cavity is packed with two coiled tubes—the small intestine and the large intestine. The small intestine, which is four times longer than the large intestine, has the job of removing and distributing **nutrients** from the thick, soupy chyme—what the food has been turned into—that squirts in through a valve leading from the stomach. As chyme enters the top of the small intestine, digestive fluids from the nearby gallbladder and pancreas flow in to help the small intestine continue the job of breaking down food into its different chemical parts.

By the time food empties into the large intestine—the next stop after the small intestine—most nutrients have been removed. The main job of the large intestine is to absorb water from the undigested food and provide a storage space for what's left of the food until it can be expelled from the body as feces. **Bacteria** that live in the large intestine feed on some of the remaining nutrients and in the process make gas, which is what gives feces such a bad odor.

Small Intestine

The small intestine's name is pretty misleading! It is actually a large **organ** with a big job to do. From the outside, the small intestine looks like what it is: a long tube about 3 to 6 cm (1 to 2 in.) in diameter coiled into a cramped space. Inside, its walls are lined with thousands of fingerlike bumps called villi, which absorb nutrients from the food and transfer them to the bloodstream. The villi are about 1 mm (0.04 in.) long, and each is coated with smaller, hairlike protrusions called microvilli. This furry lining

How Big?

provides a lot of surface area for absorbing nutrients in a very small space. It acts sort of like a fuzzy bath towel. If the walls of the small intestine were smooth, then the intestine would have to be much longer to take in as many nutrients. In fact, it would have to be about 3.5 km (2.25 mi.) long! Next time you drive somewhere with your parents, measure this distance on the odometer. You'll be amazed how long it is.

Rectum

Small Intestine

Large Intestine

Strange But True!

Tapeworms

Have you ever eaten a lot of food and then heard someone say, "You must have a tapeworm"? Of course they're kidding, but tapeworms do exist. These parasites enter the digestive tract as tiny larvae in contaminated meat or fish and can grow to be nearly 9 m (30 ft.) long. The coiled ball at left is a life-size specimen of a grown tapeworm. Although the thought of a tapeworm in your body may be disgusting, the worm is usually not a danger to its human host.

Down a Furry Tube

By the time food enters the small intestine, it has been turned into something that looks like a thick milk shake. This substance swirls through the tube, aided by **muscle** contractions and by **mucus** and digestive juices that help to move the food along. Nutrient **molecules** come into contact with the fingerlike villi *(diagram at right)* that line the tube. **Capillaries** in the villi pick up amino acids and sugars and send them into the bloodstream, while vessels called lacteals pick up fatty acids and deposit them into the lymph vessels of the **immune system** *(page 98)*. From the small intestine, blood and **lymph** go to the liver for cleansing *(pages 92-93)*.

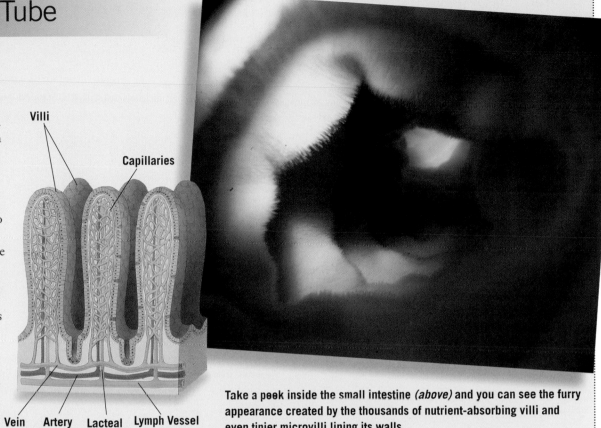

Villi

Capillaries

Vein   **Artery**   **Lacteal**   **Lymph Vessel**

Take a peek inside the small intestine *(above)* and you can see the furry appearance created by the thousands of nutrient-absorbing villi and even tinier microvilli lining its walls.

The Last Stop

By the time food gets all the way to the large intestine, only water, undigested fiber, and a few **minerals** remain. The large intestine, which measures about 1.5 m (5 ft.) long and 5 to 8 cm (2 to 3 in.) in diameter, is divided into three sections: the cecum, where the food enters from the small intestine; the colon, which is the biggest part and which absorbs most of the water; and the rectum, which stores leftover waste material. Mucus secreted from the walls of the large intestine binds together the waste into feces, which are expelled through the anus when you go to the bathroom.

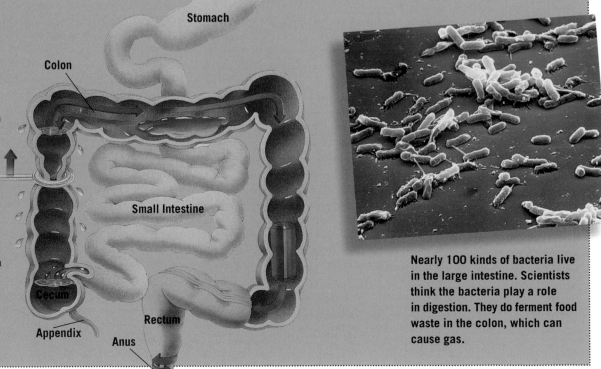

Stomach

Colon

Small Intestine

Cecum

Appendix

Rectum

Anus

Nearly 100 kinds of bacteria live in the large intestine. Scientists think the bacteria play a role in digestion. They do ferment food waste in the colon, which can cause gas.

Gallbladder, Pancreas, and Liver

Three internal **organs** vital to digestion make substances that help break down food into **nutrients.** The pancreas, a wedge-shaped **gland** located just behind the stomach, makes strong juices that help digest food—mostly **proteins**—in the small intestine. It also produces the **hormone** insulin, which helps control the level of sugar **(glucose)** in the blood. If blood-sugar levels get too high or too low, you could go into a coma.

The body's main chemical factory, the liver, lies next to the stomach. Of the hundreds of tasks it performs, most involve processing the blood that flows through it, traveling from the small intestine on its way to the heart.

Before the blood is circulated to the rest of the body, the liver cleanses it of any toxic substances that may have been swallowed along with food or drink. This important organ also stores **vitamins** and releases them into the blood when the body needs them. And it produces bile, a yellowish fluid that aids digestion in the small intestine.

The gallbladder, a sac that lies on the underside of the liver, is a storage area for bile, which breaks down the fat **molecules** in food into small pieces so the body can more easily digest them. If the gallbladder becomes diseased, it can be removed without fear of disturbing the digestive process. The bile will simply flow directly from the liver into the small intestine.

The Juice Makers

The liver is the largest organ inside the body, weighing in at about 1.4 kg (3 lb.). It produces bile, a yellow green fluid essential to digestion. It also filters toxins out of the nutrient-rich blood that is returning to the heart from the small intestine *(pages 90–91).*

The gallbladder is a muscular bag about 8 cm (3 in.) long that is tucked with-in the lobes of the liver. It stores the bile that flows into the small intestine to aid in the digestion of **fat.**

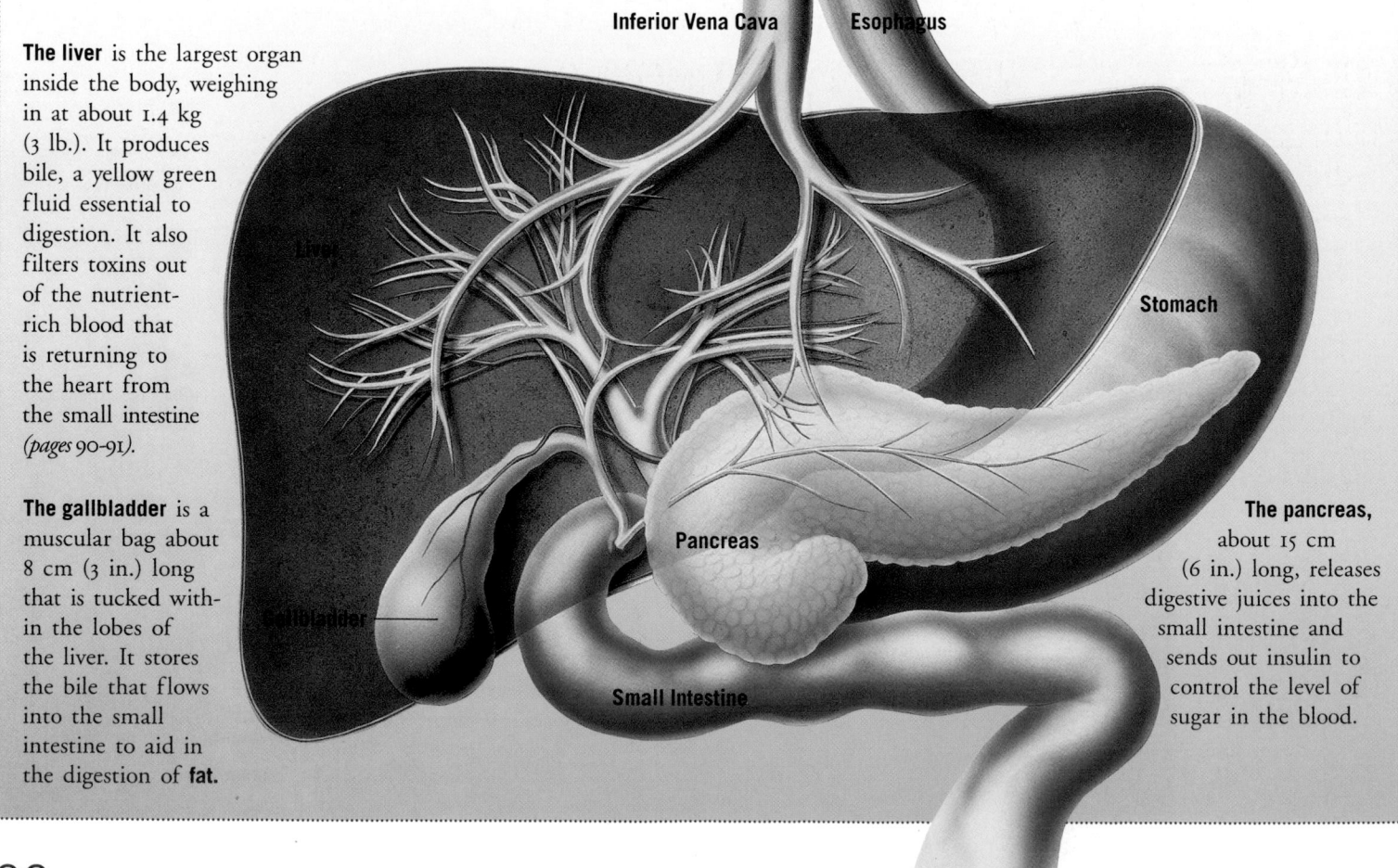

Inferior Vena Cava   Esophagus

Liver

Stomach

Gallbladder

Pancreas

Small Intestine

The pancreas, about 15 cm (6 in.) long, releases digestive juices into the small intestine and sends out insulin to control the level of sugar in the blood.

What's Diabetes?

Sometimes the pancreas doesn't produce enough insulin, or the body doesn't use the hormone as it should. When this happens, the amount of sugar in the blood becomes too high, leading to a disease called diabetes. Following a strict diet can sometimes control the condition. But many people—including children—have a type of diabetes that requires them to manage their blood-sugar levels with daily injections of insulin. Diabetic people learn to give themselves shots so they won't have to go to the doctor so often. The injections generally allow people with diabetes to go about their normal activities.

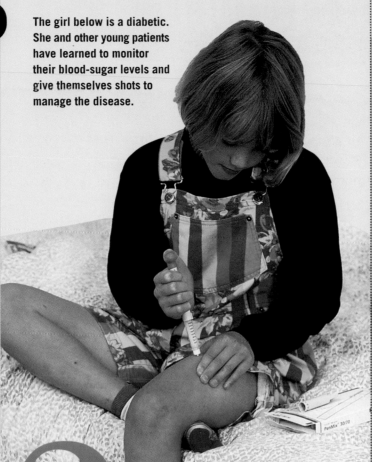

The girl below is a diabetic. She and other young patients have learned to monitor their blood-sugar levels and give themselves shots to manage the disease.

Gallstones

Bile produced in the liver contains mineral salts, which are usually dissolved in the yellowish liquid. But sometimes the salts solidify into hard lumps called gallstones. If gallstones block the tube connecting the gallbladder to the small intestine, they can cause severe pain and must be removed by surgery. Gallstones *(below)* can get quite large.

Reading the Liver

This clay model of a liver, marked with a pattern of small squares, was made in the 18th or 19th century BC. Discovered by scientists in 1890, in the ruins of the ancient city of Babylon, the object is believed to be a chart used by priests called haruspices: They claimed they could tell the future by studying the liver of a sacrificed sheep. People in some ancient cultures believed the liver was the body's physical and spiritual center.

Would You Believe?

Fast FACTS

The liver is a compact chemical plant that performs more than 500 important tasks for the body. This amazing organ looks simple, but inside it is a complex system. Without it, the body cannot survive and would die within 24 hours—and no machine has yet been invented that can step in to do its job.

The liver stores up the extra vitamins that the body takes in and releases them into the bloodstream when you need them.

Even if 90 percent of the liver were removed, the remaining 10 percent could regrow the entire organ!

The liver can filter out a remarkable range of poisons, including gold, mercury, TNT, and snake venom.

The liver's worst enemy is alcohol—including that found in beverages such as beer, wine, and whiskey. Alcohol poisons the liver and, over time, destroys the organ's **cells.**

What Is Nutrition?

Your body's basic structure—the **cells** that make up your skin, bones, **muscles,** and **organs**—is built from **nutrients** contained in what you eat and drink. To keep your body healthy and strong, you must take in a variety of foods. Nutrition is the science of determining which foods you must eat to accomplish this.

The body uses different nutrients for different purposes. Some, like **protein,** provide help for bones and muscles. Others contribute the building blocks for cells. Fiber gives food bulk as it moves through the digestive tract, allowing the intestines to massage the food and digest it thoroughly. Still other nutrients keep the body's chemical processes going, and help produce the energy needed to keep the heart beating, the lungs breathing, and the body warm. And water, which makes up nearly two-thirds of the body, helps regulate your temperature and is a necessary ingredient of both cells and blood.

A Balanced Diet

Nutritionists use charts to help people select healthful foods in the proper amounts. The table chart at right shows what proportion of your daily diet should consist of five food groups—grains; fruits and vegetables; dairy products; meat, fish, nuts, beans, and eggs; and **fats** and sugar. The largest part of your daily diet should consist of grains, fruits, and vegetables. Eat smaller amounts of dairy products, meat, and fish, and allow yourself only a little fat and sugar.

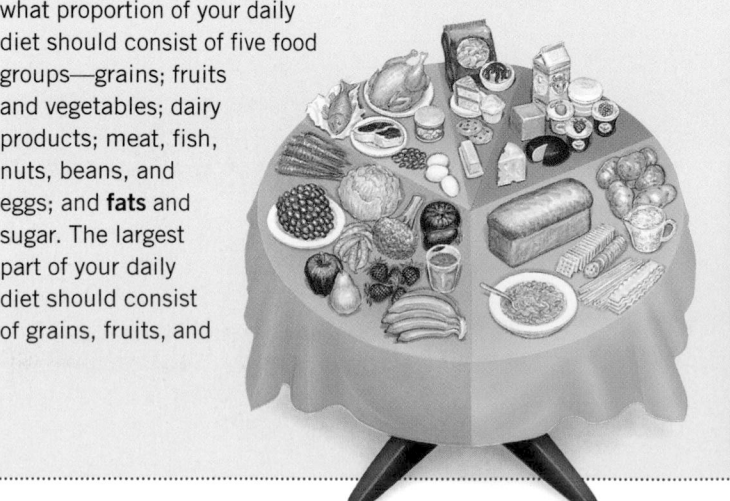

Proteins

Proteins are found primarily in meat, fish, eggs, nuts, and meat alternatives such as beans, lentils, chickpeas, and tofu. Proteins are needed by the body for the growth and repair of cells. Proteins also form parts of the skin, bones, muscles, and red blood cells. And the dozens of different **enzymes** that help to digest food are proteins.

Vitamins & Minerals

Vitamins are abundant in fruits and vegetables. In comparison with nutrients like protein, the amount of **vitamins** your body needs is fairly small, but many chemical processes depend on them. The structural upkeep of cells is the job of **minerals.** For example, strong bones and teeth depend on **calcium** —found in milk—whereas sodium, or salt, is necessary to keep nerve signals flashing around the body.

Fats

Also known as lipids, fats are found mainly in dairy products, meat, and oils. They combine with proteins to create the **membrane** around each cell, and they form the myelin sheath that insulates the axons in many nerve cells (*page 49*). Fat stored in the body's cells provides a reserve source of energy, but eating too much fat can cause problems for the heart and **circulatory system.**

Carbohydrates

These nutrients are the basis of a healthful diet. Found in starches such as bread, pasta, cereal, rice, and other grains, carbohydrates are the body's major energy-producers. In the process of digestion, most carbohydrates are broken down into sugars such as fructose and **glucose** that the body absorbs easily and quickly through the villi in the small intestine (*page 91*).

Water Use

Water makes up two-thirds of the human body and is a vital component of both cells and the blood. Water lost through normal body functions must be replaced in order for you to live.

WATER IN

Chemical reactions inside the body create some water.

Water in food provides a significant amount of fluid.

The body gets most of its water through beverages. It's important to drink water daily!

WATER OUT

Some water leaves the body in the feces.

Water evaporates through sweat.

A lot of water evaporates through your skin's pores and in breath from the lungs.

About half of the water the body gives off each day is excreted in urine.

A Diet of Worms

This South African woman is preparing a meal of protein-rich mopani worms for her family. The dish may not look appetizing to you, but to the Venda tribe the worms are an important food— and cooking them in an iron pot adds traces of that important mineral to the meal as well. Although people from different cultures may eat very different foods, all human beings need the same nutrients to keep their body healthy and strong.

The Kidneys

Just above the waist, on each side of the spine, lie your kidneys, two fist-size, bean-shaped **organs.** Their job is to filter wastes from the blood, and they're on duty all the time. In fact, every few minutes, the body's entire blood supply passes through the kidneys.

To accomplish this big job, each kidney contains more than a million tiny, blood-filtering units called nephrons. Laid end to end, the nephrons would stretch for 80 to 120 km (50 to 75 mi.). This filtration system separates harmful substances from useful ones, sending the wastes through tubes called ureters to be stored in the bladder. When you go to the bathroom, these wastes are expelled in the fluid called urine. However, most of the blood that has passed through the nephrons is reabsorbed into the **capillaries** and **veins.** Only about 1 percent of the fluid that enters the kidneys leaves the body as urine.

Inside the Kidney

Blood enters the kidney through the renal **artery** and is channeled directly to the nephrons, which are located in the cortex, or outer layer, of the kidney. (The inner layer is called the medulla.) Each nephron is composed of a cluster of capillaries, which filter the blood and then return it to the heart through the renal vein. Wastes—water, salts, and urea, a substance formed when **proteins** are broken down—are collected in the renal pelvis, then sent down the ureters to the bladder.

How *Much?*

Blood Flow

Your kidneys are full of blood vessels *(below)*. That's because they filter your entire blood supply 360 times a day. About one-quarter of the blood that's in the body—or more than 1 l (1 qt.)—moves through the kidneys each minute! If you're having trouble picturing that, imagine 1,700 l (1,800 qt.) of milk pouring through your fists. That's how much blood the kidneys process every day!

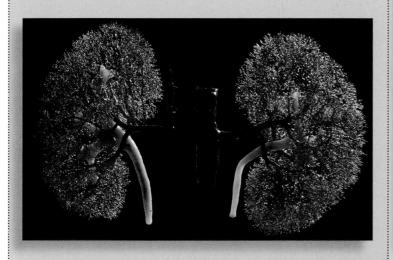

When Kidneys Fail

Sometimes kidneys can be injured or become diseased and lose the ability to filter blood. This allows harmful wastes to build up, causing illness. In such a situation, a patient's blood supply is filtered through a machine, in a process called dialysis. Another way to treat kidney failure is to transplant a healthy kidney from a relative or someone who has died. To ensure the operation's success, the kidney must be kept cool and moist until surgery *(below)*.

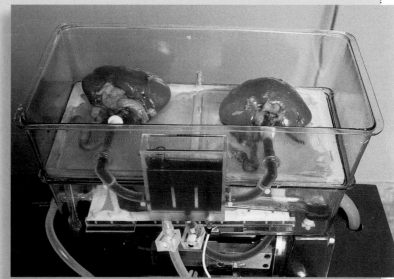

Urinalysis

Even in ancient times, people knew that urine could reveal important information about what was happening inside the body. Greek, Roman, and Arab doctors examined their patients' urine for color, smell, consistency, and taste to try to diagnose an illness. The medical chart below, from the Middle Ages, links urine color and sediment with various ailments. And today, urinalysis conducted in a medical laboratory is used to diagnose many conditions, including infections, pregnancy, and also kidney function.

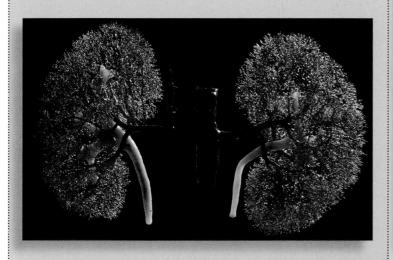

What Does Urine Contain?

Urine is made up of water and waste products—including urea, the main by-product your **cells** give off as they burn energy. If the urea stayed in your cells, it would eventually poison you. Luckily, the body produces urine as a means to wash urea and other toxins from your system. The exact composition of urine depends on what and how much you eat and drink, but generally, it is about 95 percent water.

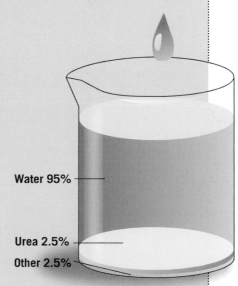

Water 95%

Urea 2.5%

Other 2.5%

The Immune System

Right now, as you read this, your body's **immune system** is protecting you against invaders—microscopic organisms such as **viruses, bacteria,** parasites, and fungi. Your skin *(pages 66-67)* and the slimy **mucus** inside your nose are part of your body's first line of defense against these attackers. But if a cold virus or other disease-causing organism gets past these defenders and enters your body, your immune system has an army of remarkable **cells** ready to continue the fight.

This army is made up mostly of different kinds of white blood cells, which are created primarily in the **bone marrow** and **spleen.** Some of the cells gobble up the invaders; others puncture or poison them. Still others produce **germ**-killing **proteins** called **antibodies** that lock on to the intruders.

The white cells usually win the battles. But sometimes they become disarmed, allowing the outsiders to take over, and possibly kill, the body they've invaded.

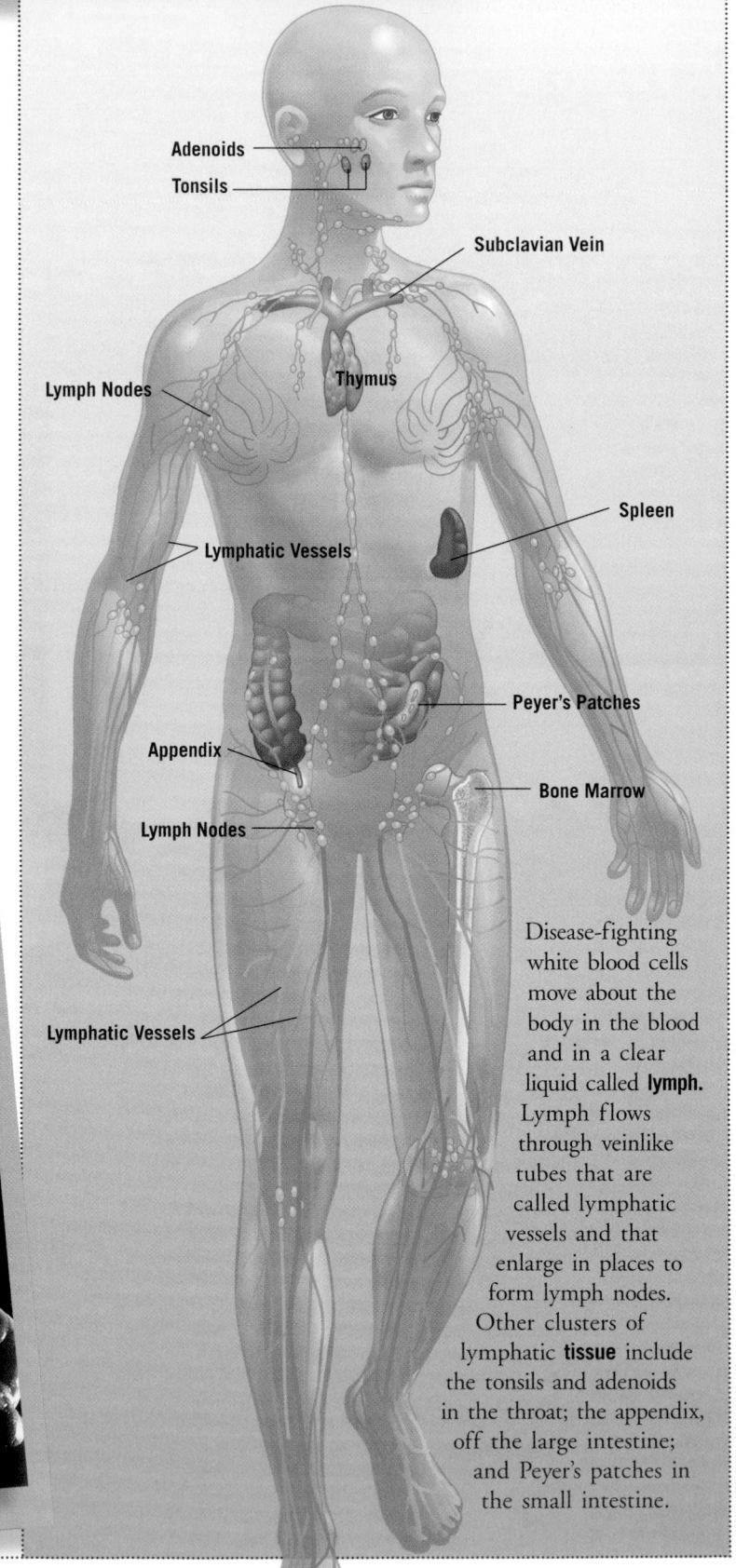

Adenoids

Tonsils

Subclavian Vein

Lymph Nodes

Thymus

Spleen

Lymphatic Vessels

Peyer's Patches

Appendix

Bone Marrow

Lymph Nodes

Lymphatic Vessels

Disease-fighting white blood cells move about the body in the blood and in a clear liquid called **lymph.** Lymph flows through veinlike tubes that are called lymphatic vessels and that enlarge in places to form lymph nodes. Other clusters of lymphatic **tissue** include the tonsils and adenoids in the throat; the appendix, off the large intestine; and Peyer's patches in the small intestine.

The Body's Battlefields

White blood cells fight infection in hundreds of bean-shaped **glands** called lymph nodes found throughout the body. You can see one *(arrow)* in tissue near the intestine pictured at right.

When the immune system is fighting a particularly fierce infection, the lymph nodes swell as they collect millions of white blood cells and dead germs. That's why you can sometimes feel sore, tender, swollen glands in your neck.

Boy in the Bubble

When David was born in 1971, he was immediately put into a plastic isolation bubble. His parents correctly suspected that he would have the same rare **genetic** disease that had killed his older brother. David's immune system could not fight off germs. Any exposure to a germ—even a cold virus— would kill him.

To stay alive, David had to live in a sterile, or germ-free, environment. A four-room bubble was built for him in his home in Texas. His toys, clothes, and food were sterilized before being put in the bubble. No- body, not even his parents, could enter the bubble for fear of bringing in germs.

When David was 12, he underwent an experimental operation to give his body the white blood cells it needed. Sadly, the operation failed. David died in 1984, 15 days after leaving his bubble.

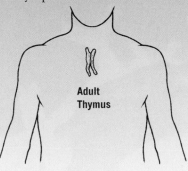

David, shown above with a nurse, lived a full life in his plastic bubble. He attended school by speakerphone, watched television, and played with friends using arm-length tubes attached to his bubble *(left)*. When David was six, NASA gave him a spacesuit so he could sometimes play outside.

What Are Tonsils For?

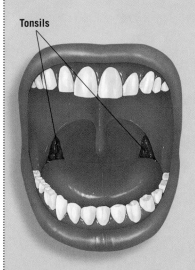

Tonsils

Your tonsils—those two oval- shaped globs of tissue at either side of the back of your mouth—help keep bacteria and other organisms from entering your throat. They also help your body make disease-fighting white blood cells and antibodies.

Sometimes viruses or bacteria inflame the tonsils, causing them to swell and become painful. This inflammation is called tonsillitis. Doctors used to routinely remove sore tonsils, but they now prefer to treat the infection with drugs.

Let's **Compare**

Thymus Glands

The thymus is a mysterious growing and shrinking gland. Located in your upper chest, just behind the breastbone, it is fairly large at birth. From then until puberty, the thymus is an important source of lymphocytes— white blood cells that are part of the immune system. By puberty, the thymus is about 8 cm (3 in.) long. Then it begins to shrink, wasting away until almost nothing is left of it by old age. By this time, the manufacture of lymphocytes has been taken over by the bone marrow and lymph nodes.

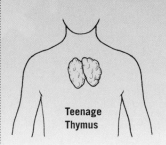

Teenage
Thymus

Adult
Thymus

Fighting Disease

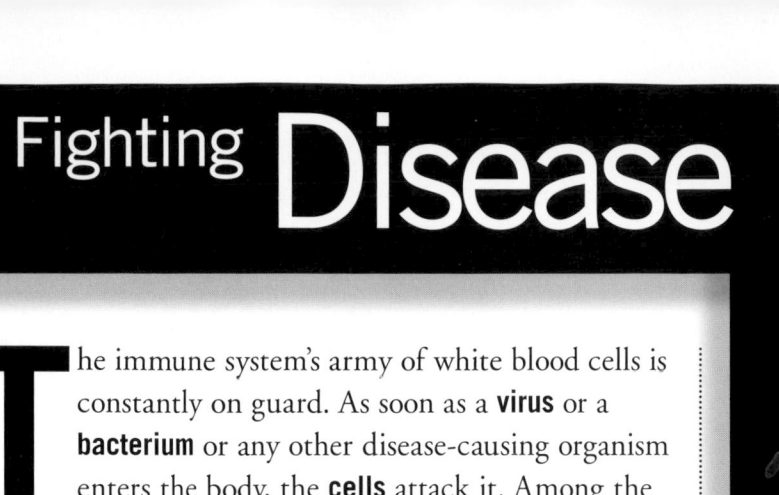

The immune system's army of white blood cells is constantly on guard. As soon as a **virus** or a **bacterium** or any other disease-causing organism enters the body, the **cells** attack it. Among the first on the scene are the **macrophages** *(right)*. These cells regularly patrol the body, devouring **germs** and other potentially dangerous debris—like the dust that collects in your lungs. They also set off a chemical alarm that tells other white blood cells to prepare for battle.

Coming quickly to the rescue are killer T cells *(opposite)* and B cells *(below)*. Killer T cells destroy those cells in the body that the invading organism has already successfully infected. B cells attack the organism itself by producing **proteins** called **antibodies.**

It can take the **immune system** a few days to get its army of white blood cells fighting at full strength. You may feel sick until the invading germs are gone.

A macrophage uses its long, ropelike pseudopods to grab *E. coli* bacteria *(green)*. The macrophage will gobble up the captured bacteria and then search for more. At top left a red blood cell floats by.

B Cells Go to Battle

B cells make their own germ-fighting weapons, called antibodies. The antibodies extend outward from the B cells like tiny Y-shaped antennae. There are more than a million different types of antibodies. Each one is shaped to match the specific marker **molecule,** or **antigen,** of an invading organism.

Antigen

Invading Organism

New B Cells

B Cell

Antibody

When a B cell comes into contact with an antigen for which it has an antibody, it locks on to it and begins dividing. Soon, millions of new B cells have been created.

When enough new B cells have been made, most of them stop dividing. They become plasma cells, a type of cell that makes free antibodies. Free antibodies don't stay attached to the B cells, but roam about freely. They look for invading antigens that they can bind on to.

After an antibody attaches itself to a target antigen, the antibody changes into a shape that is easier for macrophages to latch on to. By releasing antibodies, therefore, B cells help macrophages devour and kill invading organisms.

Macrophage

Some B cells continue dividing for a long time, sometimes for years. If a germ tries to invade the body again, these B cells, called memory cells, "remember" it and are ready with the needed antibodies. That's how the body becomes immune to a disease it had in the past.

Free Antibodies

Plasma Cell

Memory Cells

A macrophage *(right)* engulfs a bunch of tuberculosis bacteria *(green)*. After the macrophage digests the bacteria, it will release a chemical that alerts other white blood cells that an invasion of tuberculosis has begun.

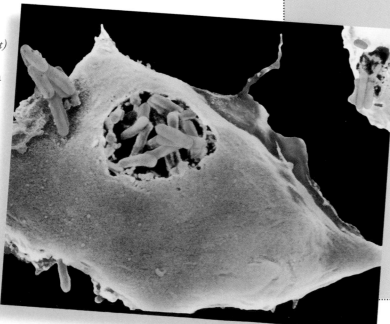

A Killer T Cell at Work

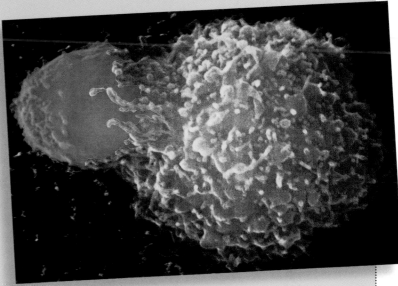

The photograph above, taken through an electron microscope, shows a killer T cell *(left)* beginning its attack on a **cancer** cell. Killer T cells are foot soldiers in our body's defense system. They seek out and kill invaders by injecting them with a chemical that eventually causes cell death.

What's in a

Name?

Autoimmune Disease

Sometimes the immune system loses its ability to tell a friend from an enemy. When that happens, it begins to attack the body's own **tissues.** No one knows just why this occurs; in some cases, it may start when the body is invaded by a virus. This "autoimmune" attack can lead to different kinds of diseases, such as Graves' disease, systemic lupus, and myasthenia gravis. The "auto" in autoimmune, which comes from the Greek word *autos,* means "acting on oneself."

Building Immunity

S mallpox used to be a deadly disease, feared by people around the world. During the 18th century, nearly 60 million Europeans alone died from the smallpox **virus**. As late as 1967, smallpox was killing about two million people each year.

This year, thanks to a successful worldwide vaccination campaign, no one will die of smallpox. The last known natural case of smallpox occurred in Somalia in 1977. This once terrifying disease has disappeared from the face of the earth.

Vaccines like that for smallpox are among medicine's most exciting success stories. By injecting a dead or harmless form of the disease into the body, **vaccines** trick the **immune system** into thinking the body has been invaded by a disease. The immune system goes through the process of destroying the invader. It also prepares itself to recognize the invader again. This gives the body **immunity**—long-term protection—against the disease.

Fast FACTS

Vaccines

Smallpox Introduced: 1798. Annual cases before vaccine: several million worldwide. After: none after 1977.

Tetanus Introduced: 1938. Annual cases before vaccine: 500 to 600 in U.S. After: 50 to 100.

Whooping cough (pertussis) Introduced: mid-1940s. Annual cases before vaccine: 150,000 to 260,000 in U.S. After: 2,000 to 8,000.

Polio Introduced: 1954. Annual cases before vaccine: 10,000 to 50,000 in U.S. After: fewer than 1,000.

Measles Introduced: 1963. Annual cases before vaccine: three million to four million in U.S. After: 150,000 to 200,000.

Mumps Introduced: 1968. Annual cases before vaccine: 150,000 in U.S. After: 1,000 to 6,000.

Hemophilus influenzae b (Hib) Introduced: 1987. Annual cases before vaccine: 20,000 in U.S. After: 200 to 600.

Famous 1 FIRSTS

Vaccination

W hen Edward Jenner was a teenager in England during the 1760s, he heard a young farm girl claim that she could not get smallpox. She said she had already had cowpox, a similar but less deadly form of the disease. The girl's statement got Jenner thinking: What if it were true that having cowpox protected people against smallpox? Thousands of lives could be saved each year.

Jenner became a doctor. For the next 20 years he experimented with cowpox to find out if it provided protection, or immunity, against smallpox. In 1796 he was ready to test his theory. He scratched the pus from a cowpox sore into the arm of a healthy eight-year-old boy, James Phipps (*right*). The boy came down with cowpox, but he recovered quickly. Then Jenner scratched the boy's arm with smallpox. James did not get ill.

Jenner called his new procedure "vaccination." He had shown for the first time that the body could "remember" certain diseases and fight them off when they returned.

Shots Save Lives

Children, like this little girl in Haiti, are understandably nervous about getting their immunization shots. But such shots have saved millions of lives. That's why the world's health organizations are reaching out to developing countries in an effort to get as many children vaccinated as possible.

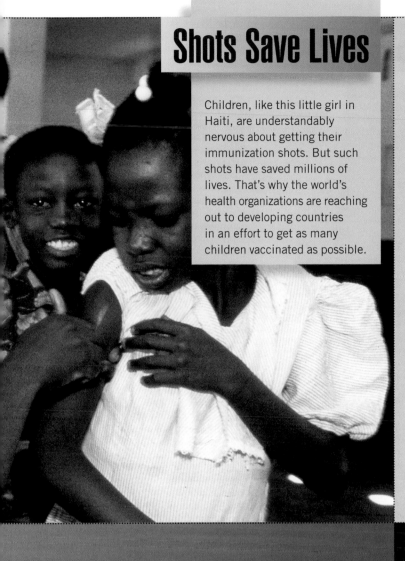

Laugh Yourself Healthy

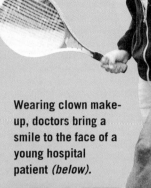

In 1964, Norman Cousins *(right)*, an American magazine editor, came down with an often deadly form of arthritis. He was given a 1 in 500 chance of survival. But Cousins wasn't about to give up. He decided that he would help his body fight the illness by keeping a positive attitude—and by laughing as much as possible.

Wearing clown make-up, doctors bring a smile to the face of a young hospital patient *(below)*.

Each day, Cousins read funny books and got lots of laughs watching films starring the Marx Brothers. Much to his doctors' surprise, Cousins recovered from his illness. Scientists have since found that laughter does seem to help the body's immune **cells** fight disease.

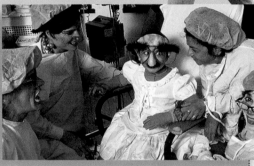

People   Jonas Salk

Jonas Salk (1914-1995), an American doctor and researcher, developed the first vaccine against polio. This dreaded disease used to kill and cripple thousands of people, especially children, each year.

Salk figured out how to take the live polio virus, kill it, and then inject it back into the body as a vaccine *(right)*. The polio vaccine was first widely distributed in 1955. By 1962, fewer than 1,000 new cases of polio were reported in the United States—down from about 50,000 ten years earlier!

Bacteria and Viruses

Wash Your Hands!

When you get sick, it's probably because of either a **virus** or a **bacterium.** These are the two most common kinds of pathogens, or disease-causing organisms.

Bacteria are living, one-celled organisms that can be seen only under a microscope. They come in many shapes, from rods to spirals. Your body contains about 100,000 billion bacteria. Fortunately, most are harmless. Some actually help your body—by digesting your food, for example *(page 91).* But others can make chemical poisons that destroy human **tissue.** Diseases such as pneumonia, strep throat, tuberculosis, and some kinds of food poisoning are caused by bacteria.

Viruses are extremely small particles that cannot live on their own. They multiply only when they enter the **cell** of another organism. Once inside, a virus produces new viruses until the host cell can no longer hold them and bursts. The common cold and flu, measles, rabies, and AIDS are some of the diseases caused by viruses.

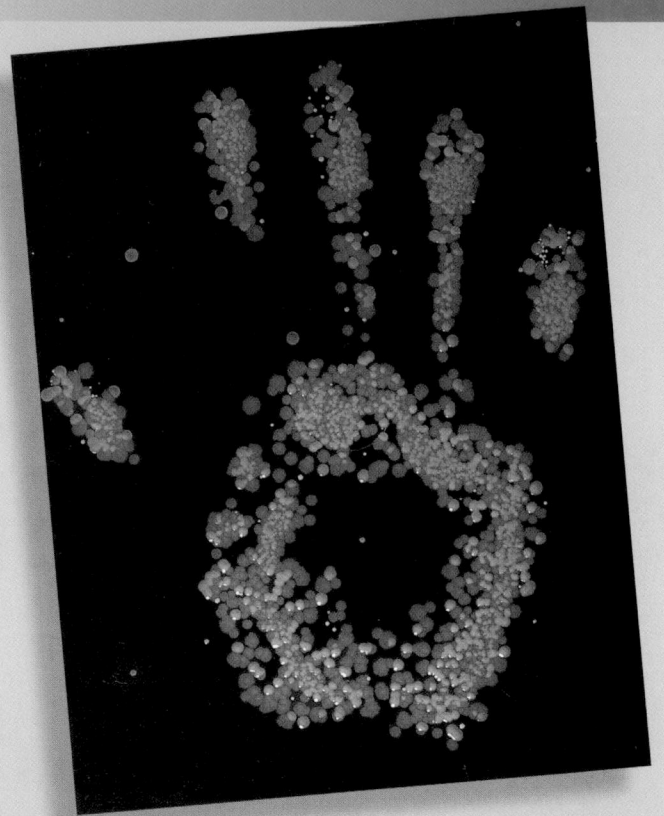

Why should you wash your hands before eating? Take a look at this hand print. It was made by pressing the palm of a hand onto a special surface that allows bacteria to grow. The red dots show where bacteria from the hand have grown and multiplied. A good hand wash would get rid of 80 percent of these germs.

Let's Compare

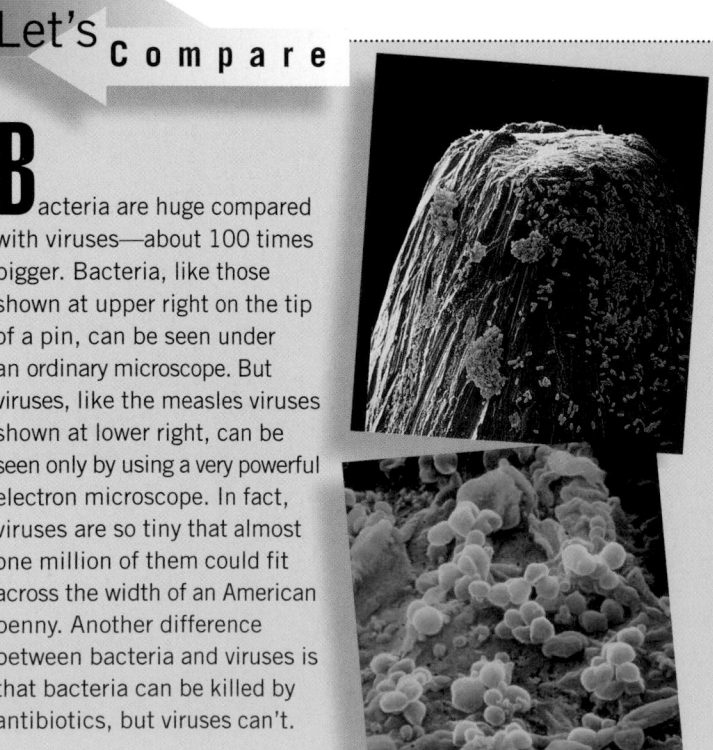

Bacteria are huge compared with viruses—about 100 times bigger. Bacteria, like those shown at upper right on the tip of a pin, can be seen under an ordinary microscope. But viruses, like the measles viruses shown at lower right, can be seen only by using a very powerful electron microscope. In fact, viruses are so tiny that almost one million of them could fit across the width of an American penny. Another difference between bacteria and viruses is that bacteria can be killed by antibiotics, but viruses can't.

People — Louis Pasteur

The brilliant French chemist Louis Pasteur (1822-1895) was one of the first scientists to believe that bacteria and other tiny organisms cause infectious diseases. He proved, for example, that bacteria spoil milk and developed ways of killing them. This process was named pasteurization after him.

Pasteur also developed the **vaccine** for rabies, a viral disease. His first shots saved the life of a nine-year-old boy who had been bitten by a rabid dog.

How Germs Are Spread

Bacteria and viruses can spread from one person to another. When you have a cold and cough or sneeze, you blow out millions of tiny droplets of **mucus.** Each droplet carries the cold virus. Other people may breathe them in or pick them up on their hands.

Flies and cockroaches also spread **germs** by poisoning food with bacteria picked up from dirty places. In tropical lands, mosquitoes sometimes spread deadly viruses like yellow fever and encephalitis. The mosquito picks up the virus by biting someone who has the disease *(below).* It then passes the germ to the next person it bites.

People have always been scared of epidemics—mass outbreaks of disease that sweep across cities, countries, and sometimes even continents. They have had good reason to be fearful. An epidemic of the bubonic plague during the 14th century killed 25 million people in Europe alone, or about one-fourth of that continent's population. During 1918-1919, right after World War I, an epidemic of a particularly deadly flu swept around the world. It killed 20 million people, more than twice as many as died in the war. Fortunately, the discovery of vaccines and drugs like penicillin has helped make it more difficult for infectious diseases to spread.

Flu struck so many people in 1918 that doctors had to set up tent hospitals, such as this one in Massachusetts, to care for all of them.

d Casualties this Week.

Impofthume	11
Infants	16
Killed by a fall from the Belfrey at Alhallows the Great	1
Kingfevil	2
Lethargy	1
Palfie	1
Plague	7165
Rickets	17
Rifing of the Lights	11

As this old death list shows, more than 7,000 people died in London from the bubonic plague during a single week in 1665.

Penicillin: Miracle Drug

One day in 1928, Scottish biologist Alexander Fleming noticed that a green mold had accidentally invaded a dish of bacteria he was studying. The mold, called *Penicillium notatum,* had stopped the bacteria from growing.

That's how penicillin was discovered. This miracle drug has probably saved more lives than any other medicine in history. It cures many common diseases, such as strep throat and otitis media (middle ear infections).

Penicillin is an antibiotic, a substance that kills bacteria. It can turn a bacterium *(top)* into an empty shell *(bottom).*

Antibiotics are useless against viruses, such as the ones that cause the common cold and the flu.

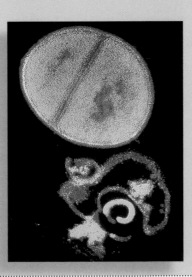

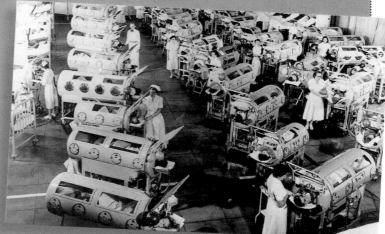

The polio epidemics of the 1940s and 1950s paralyzed or killed tens of thousands of people. These polio patients are being kept alive in breathing machines called iron lungs.

What Are Allergies?

Causes of Allergies

T he **immune system** is the body's protector, always on guard against invading **germs**. But sometimes it mistakes an ordinary, harmless substance, such as pollen or peanuts, for a dangerous **bacterium** or **virus**. This kind of immune system mistake is called an allergy.

For some people, for example, allergies are triggered by grass pollen. When such a person breathes in or touches grass pollen, the immune system identifies it as a dangerous invader and creates **antibodies** against it. The second time the person runs into the pollen, and anytime thereafter, certain **cells** release a chemical called histamine. Histamine widens the blood vessels, enabling the immune cells to rush to the rescue. Unfortunately, it also causes unpleasant allergic reactions: a runny nose, watery eyes, wheezing, and itchy skin.

Life-Threatening Allergies

A llergies can make you miserable, but they usually aren't fatal. However, some people have very serious, even life-threatening, reactions to certain foods, medicines, or insect venom. The most serious reaction, and luckily the rarest, is called anaphylaxis. It can cause **blood pressure** to drop suddenly and breathing passages to close, and it can even trigger a heart attack. Without emergency medical attention and strong drugs, anaphylaxis can be deadly. Peanuts *(top right)* and shrimp *(bottom right)* are among the most common causes of severe food allergies.

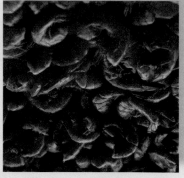

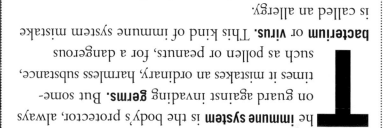

If something that you touch, eat, or breathe makes you sniffle or itch, you may be allergic to it. More than 40 million Americans have allergies.

Anything that causes an allergic reaction is called an allergen. Allergens can be found everywhere: in the air, in foods, medicines, and insect venom, even in fabrics and metals. The most common allergen is pollen *(near left)*, the tiny, powdery grains that are released by trees, flowers, and grasses and carried along by the wind.

The extreme closeup of dog hair at far left shows tiny flakes of dead skin, called dander. Dander is a common allergen that comes from the hair, skin, and feathers of animals. Like pollen, dander is small enough to float into the mouth and nose.

If dust makes you sneeze, chances are it's actually dust mites *(page 67)* that cause your allergies. Cockroaches *(above)* leave droppings whose **proteins** also cause allergic reactions.

Chemical Sensitivity

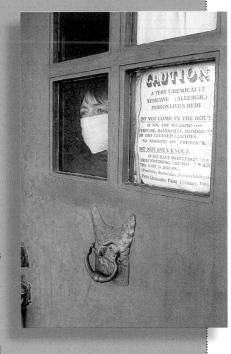

From soap to shampoo, and from plastics to paint, chemicals can be found in hundreds of things we use every day. To some people, however, chemicals can be a source of misery. For these people a single whiff of perfume or the fumes from a new carpet might bring on a headache or an asthma attack. For others, wearing clothes washed in a strong detergent might cause a red, itchy rash.

Josephine Hughes *(right)* is sensitive to so many chemicals that she will not leave her house. A sign on her door lists restrictions for visitors. Doctors disagree about the cause of chemical sensitivity. Some think it might be related to feelings of depression.

What's Asthma?

Lungs under Attack

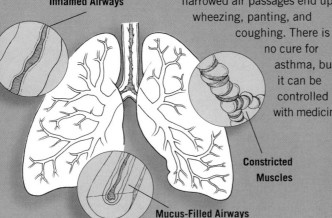

Inflamed Airways

Constricted Muscles

Mucus-Filled Airways

Asthma is an allergic reaction that makes it hard to breathe. In asthma, the airways in the lungs become inflamed and clogged with thick **mucus,** and the **muscles** surrounding them tighten. Asthmatic people who try to breathe through these narrowed air passages end up wheezing, panting, and coughing. There is no cure for asthma, but it can be controlled with medicine.

People Jackie Joyner-Kersee

Asthma doesn't have to keep you from achieving your dreams. Athlete Jackie Joyner-Kersee was diagnosed with chronic asthma when she was a 21-year-old college student. In addition to attending classes, she was training six hours a day for the 1984 Olympic Games. Sometimes, after a strenuous workout, asthma attacks left her gasping for breath. Instead of giving up, she began a program of regular medication and rest. On smoggy days she trained indoors. She went on to win three gold medals in two Olympic Games and is considered one of the world's greatest athletes.

AIDS Viral Killer

Until the 1980s, no one had heard of AIDS. Today it is well known as the cause of one of the most deadly epidemics of modern times. Scientists estimate that by the year 2000 some 40 million people will have become infected with the fatal disease.

AIDS is the acronym for acquired immune deficiency syndrome, an infectious disease caused by a **virus.** The virus is called HIV, for human immunodeficiency virus. The virus attacks the **immune system** itself, leaving the body unable to fight other infections.

Unlike the virus that causes the common cold, the AIDS virus is not spread through the air. You cannot "catch" it by being near an infected person, or by kissing or sharing food. The only known ways that people get HIV are from blood-to-blood contact or sexual contact.

Since the AIDS epidemic was first identified in 1981, scientists have been searching for a cure. None has been found yet, but thanks to new drug treatments, people with AIDS are able to live longer with the disease.

How AIDS Infects

The AIDS virus, HIV, targets a particular white blood **cell** called a helper T cell. These cells are part of the immune system and help other white blood cells fight disease.

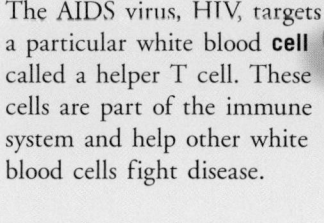
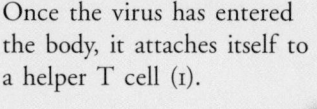

Once the virus has entered the body, it attaches itself to a helper T cell (1).

The virus then penetrates the **nucleus** of the cell (2). Once inside the cell, the virus creates new AIDS viruses.

Each new virus breaks out of the helper T cell to search for its own cell to invade (3). After it has been invaded, the helper T cell dies. Over a period of many years, enough helper T cells die that the immune system is no longer able to successfully protect the body from other infections.

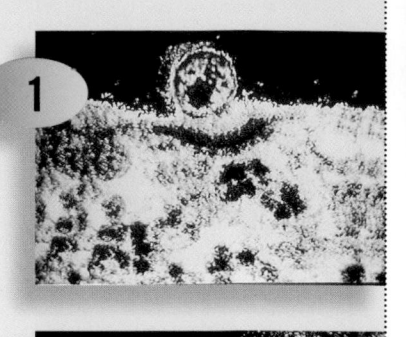

Who Discovered AIDS?

During the late 1970s doctors began noticing that an unusual number of patients were dying from a rare skin **cancer** and a form of pneumonia that hadn't been deadly in the past. The doctors called this collection of symptoms AIDS. Scientists around the world, including Robert Gallo of the United States and Luc Montagnier of France, began looking for the virus that they suspected was causing it. Both Gallo and Montagnier discovered the virus—HIV—about the same time in 1983-1984. Their discovery led to a valuable blood test for HIV.

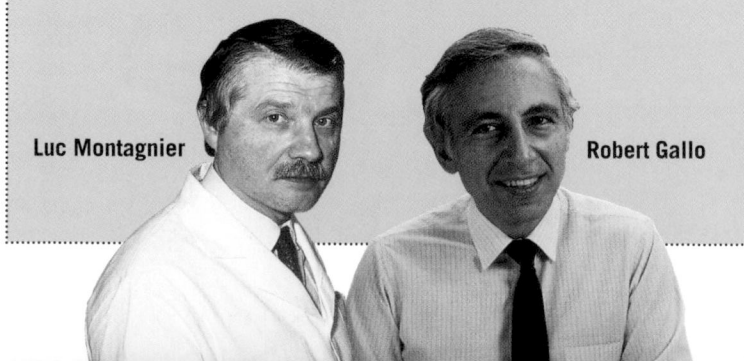

Luc Montagnier   **Robert Gallo**

The AIDS Quilt

The human side of the AIDS epidemic can be seen in the AIDS Memorial Quilt, the largest ongoing community arts project in the world. The quilt is made up of thousands of homemade 0.9-by-1.8-m (3-by-6-ft.) panels. Each panel celebrates the life of someone who has died from AIDS and is made by that person's loved ones. By 1996, the quilt was so large that it covered the entire Capitol Mall in Washington, D.C. *(right).*

The Spread of AIDS

No one knows for sure where the AIDS virus came from, but many scientists believe it was first transferred to humans from monkeys in Africa, possibly in the late 1940s or early 1950s. The virus gradually spread around the world, carried by people traveling from Africa to other countries.

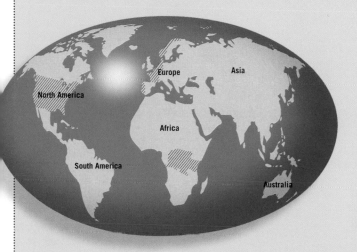

1981

By 1981, AIDS cases had been reported in the U.S., Haiti, Europe, and Africa. Several hundred people had died from the disease.

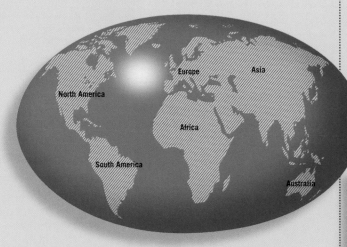

1998

By 1998, AIDS had spread to almost every country in the world. The death toll was now up to more than 11 million.

People

People Ryan White

In 1984, 13-year-old Ryan White learned he had AIDS. He got it through tainted blood he had received for his hemophilia, an illness he had been born with.

This was awful news. But Ryan decided to live a normal life for as long as possible.

When they heard Ryan had AIDS, many of the people in his Indiana hometown became frightened and angry. They refused to let him return to school. They were ignorant about AIDS and thought he would infect the other kids. Someone even fired a bullet into his house.

With great courage, Ryan began speaking out against people's misunderstandings about the disease. He became famous. Eventually, he returned to school, although to a different one. Ryan died in 1990 when he was 18 years old.

Deadly New Viruses

AIDS is only one of more than 30 new infectious diseases that have appeared in the last 20 years. One of the most deadly of the new diseases is Ebola hemorrhagic fever. It is named after the river in the Democratic Republic of the Congo (formerly Zaire), Africa, where the disease was first discovered in 1976.

Like AIDS, Ebola is caused by a virus *(below)*. Once the virus has invaded the body, it destroys the blood's ability to clot. Victims start bleeding from all parts of their body and usually die. Since 1976, outbreaks of Ebola have killed more than 600 people.

Reproductive System

The **reproductive system** is a remarkable group of **organs** that work together to **conceive** and give birth to children. Male and female bodies have different reproductive organs. The female reproductive system includes the breasts, which make milk to feed a baby. **Ovaries** house eggs. Leading from ovaries are the **fallopian tubes,** which open into the **uterus,** or womb, a muscular pear-shaped organ in which a **fertilized** egg can develop into a baby. At the bottom of the uterus is the **cervix,** which leads into the vagina.

The major organ of the male reproductive system is the penis with its rounded head, or glans. Within the penis is the urethra, a tube that carries urine outside the body, but at other times carries semen, a fluid containing **sperm,** the male sex **cells.** Sperm are produced in the testes, two **glands** that hang below the body in a skin sac called the scrotum.

In both sexes, the activity of the reproductive system is guided by the **pituitary gland** and the **hypothalamus,** both found in the brain *(pages* 50-53*)*. These glands release **hormones** *(page* 64*)*, chemicals that control sex cell development, **puberty,** and growth.

Venus and Mars

You may have heard the expression "Women are from Venus, men are from Mars." Although we aren't from different planets, the symbols for male and female *(right)* arise from the Greek symbols for Ares, the god of war, and Aphrodite, the goddess of love, beauty, and fertility. They were considered complementary opposites. In the Roman empire, the gods—and planets—were renamed Mars and Venus.

Female

Male

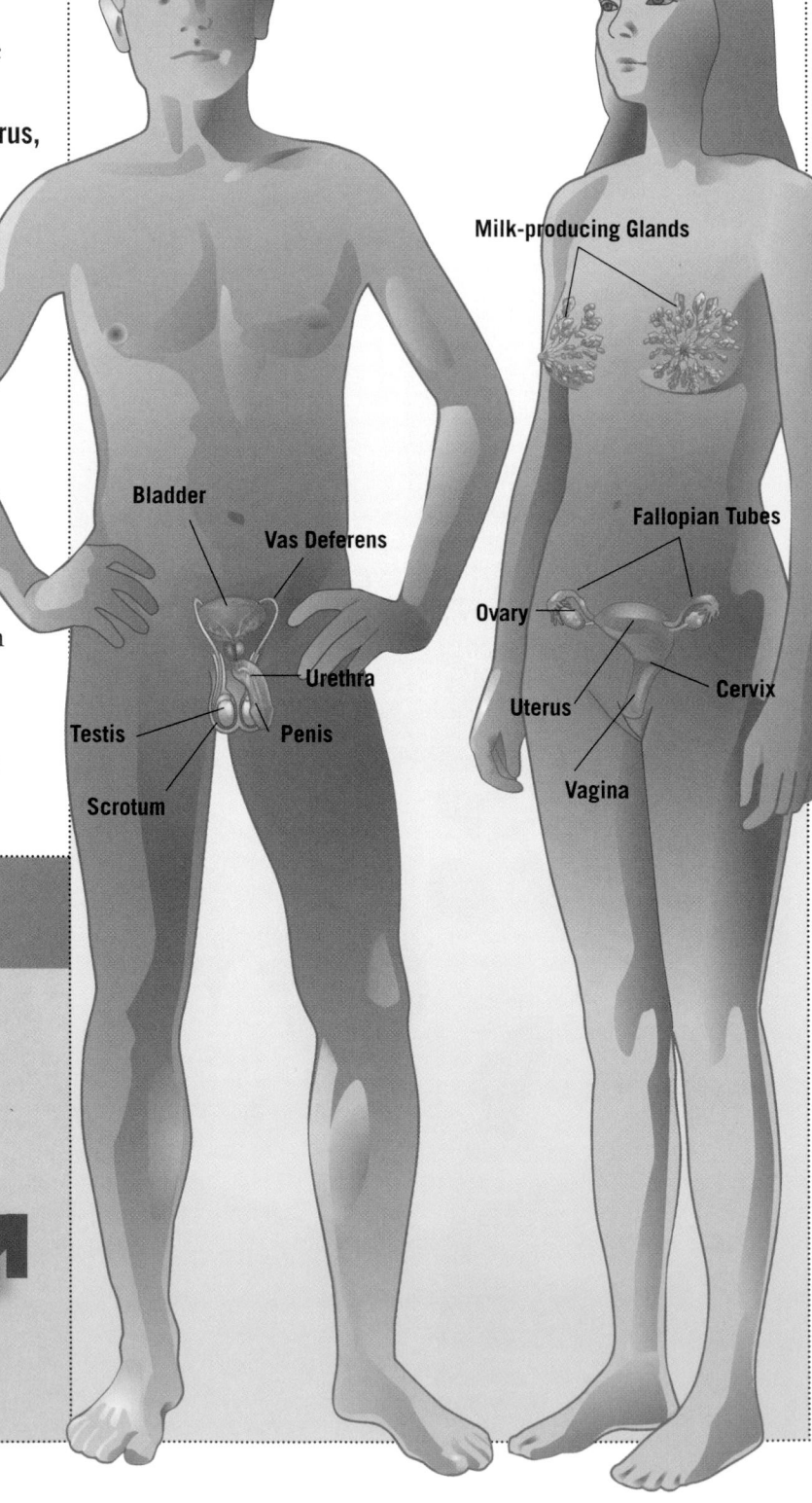

Milk-producing Glands

Bladder

Vas Deferens

Fallopian Tubes

Ovary

Urethra

Uterus

Cervix

Testis

Penis

Vagina

Scrotum

The Sex Cells

Egg

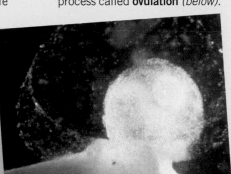

Ovary

At birth, every female has between 400,000 and two million eggs, or **ova**, in her ovaries—far more than she will ever need! A woman's body releases only about 500 ova during her reproductive life span. After birth, the female body cannot "make" any more eggs.

An ovum is about the size of a pinhead, which makes it the largest cell in the human body. It is the only cell that can be seen without a microscope.

Every month after puberty, one egg matures within an ovarian follicle, a sac that is located within the ovary. Midway through the cycle, the follicle bursts, releasing the egg in a process called **ovulation** (below).

Sperm

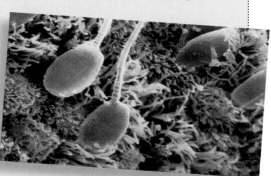

Testes

Tadpole-shaped sperm (below, right) are the smallest cells in the body—5,000 laid out end to end would measure only about 2 cm (1 in.)! A sperm's head contains **genetic** information, and its tail allows it to move. Sperm develop and are stored within the testes. An adult man makes more than 100 million sperm a day, a process that occurs best at 3 to 5 degrees below body temperature. That's why the testes hang outside the body in the scrotum. When the body is hot, the scrotal **muscles** relax and the testes lie away from the body in a cool position. When it's cold, the muscles contract, bringing the testes close to the body.

What's **Menstruation?**

Menstruation is the name for a monthly cycle that women begin to experience at puberty. Commonly known as a "period," this cycle is controlled by hormones that come from the ovaries, and from the pituitary gland and the hypothalamus of the brain. The cycle starts with ovulation—the release of an egg from an ovary. The uterus prepares itself for the egg, just in case it gets fertilized (joined with a sperm). The uterine lining thickens into a spongy cushion of blood-filled **tissue** that can nourish the fertilized egg. If the egg is not fertilized, it will fall apart and leave the body along with the uterine lining. This lining sheds slowly, dribbling out of the vagina for the next two to eight days. **Menstruation** occurs every 26 to 30 days over a woman's life. A girl usually gets her first period between the ages of eight and 17, and her last one in her 40s or 50s.

Hermaphrodite

There are animals that have both male and female sex organs. These animals are called hermaphrodites, a word that comes from the Greek god Hermaphroditus, who was part male and part female. Animal hermaphrodites include worms, snails, and some tropical fish, such as the parrotfish (below).

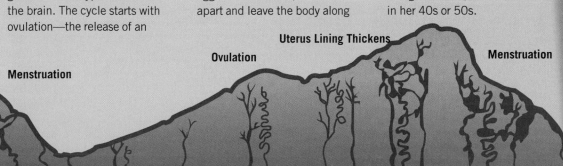

Menstruation

Ovulation

Uterus Lining Thickens

Menstruation

1 2 3 4 5 6 7 8 9 10 11 12 13 14 15 16 17 18 19 20 21 22 23 24 25 26 27 28

Fertilization

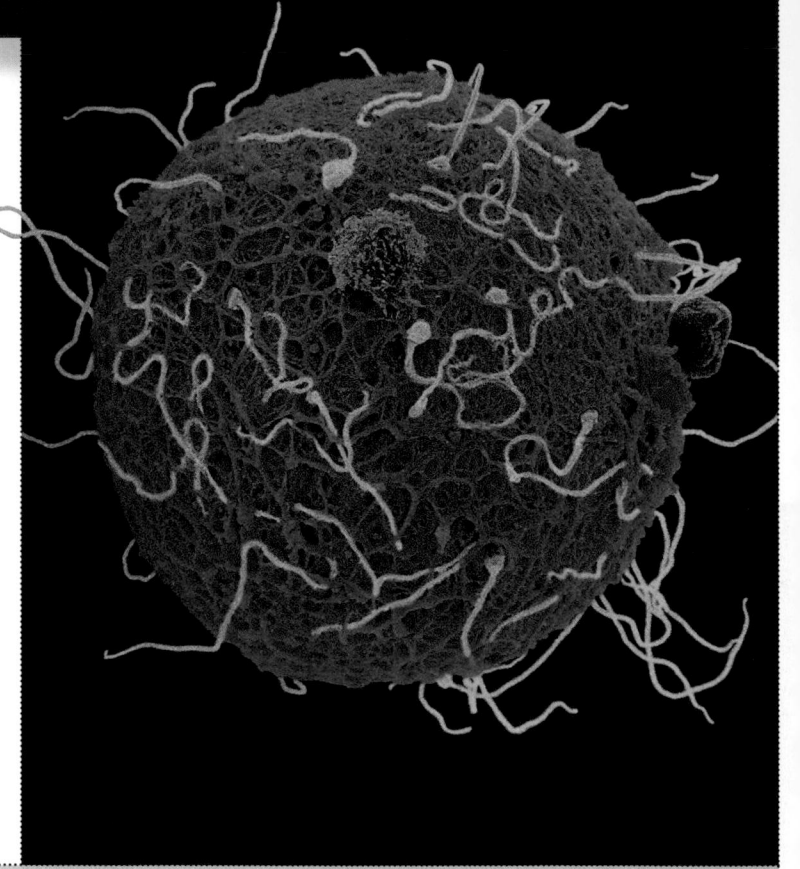

Fertilization occurs when a **sperm** and an egg join together after a man and woman have sexual intercourse. During intercourse, millions of sperm from the man are deposited in the woman's vagina. Swimming vigorously, some of the tiny sperm enter through the **cervix** into the **uterus** and **fallopian tubes.** Sperm can live in a woman two to five days. A few hundred of them will make it all the way to the egg, usually within a day.

Crowding around the egg, the sperm chemically break down its outer **membrane,** until a single sperm enters. Changes in the egg's membrane stop other sperm from entering *(right).* The successful sperm loses its tail, and its head joins together with the egg's **nucleus.** A **fertilized** egg, called a zygote, is formed. Within a week, the zygote implants itself into the wall of the uterus and begins to grow into a baby to be.

In the Beginning

Fertilization usually happens in the fallopian tube, near the **ovary.** The fertilized egg, or zygote, divides as it moves down the tube. The zygote reaches the uterus about three days after **ovulation.** There it develops into a fluid-filled capsule holding an inner bunch of **cells.** The mass plants itself in the spongy, blood-engorged uterine wall. This is the beginning of a baby.

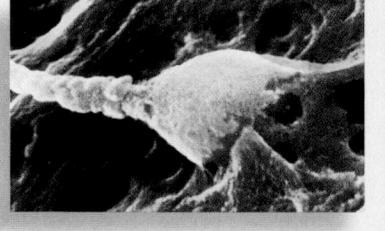

Conception magnified: A sperm enters the egg.

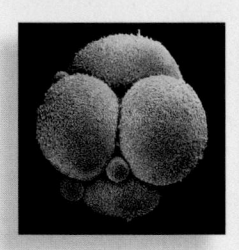 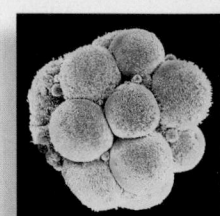

The fertilized egg begins to undergo mitosis, or cell division. The single cell splits into two, four, then eight cells *(above, left to right).*

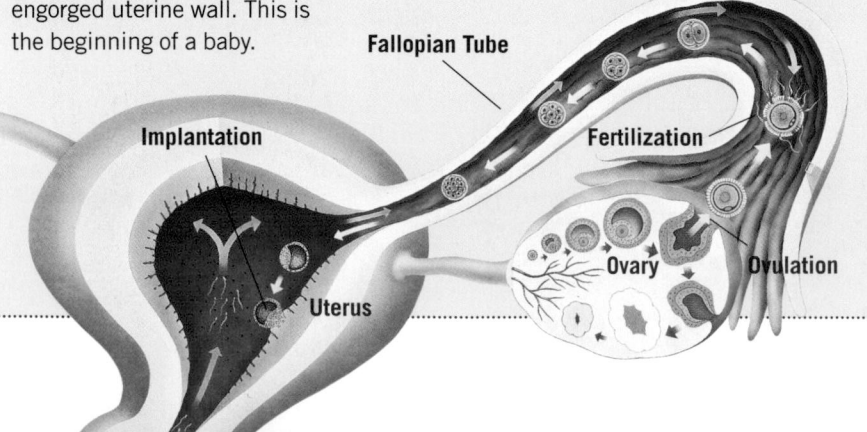

Fallopian Tube

Implantation

Fertilization

Ovary

Ovulation

Uterus

By the time the divided egg plants itself in the wall of the uterus *(right),* it is a knobby ball of more than 100 cells.

...Do Sperm Swim?

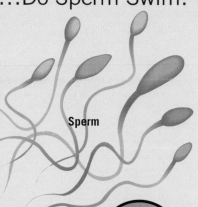

By lashing their tails, sperm can move 7.6 cm (3 in.) per hour. It takes them about 90 minutes to journey the few inches from the vagina to the fallopian tube. Not every sperm is equally fast. Scientists think that sperm carrying Y **chromosomes** swim faster, because the Y chromosome is lighter than the X *(pages 116-117).*

Sperm

Test-Tube Tyke

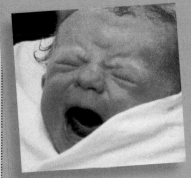

Famous **1** FIRSTS

Born in 1978 in Bristol, England, Louise Brown was the world's first "test-tube baby." Although she did not develop in a test tube, Louise was **conceived** in one. Doctors removed an egg from one of her mother's ovaries, fertilized it in a laboratory with sperm from her father, and implanted the **embryo** into her mother's uterus. This process, called in vitro fertilization, is now a common way doctors help couples have a baby.

Would **You** Believe?

Unidentical Fingerprints

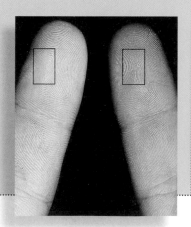

Even identical twins have one **trait** that is different—their fingerprints! As you can see in the boxed areas of this photo, twins have different ridge patterns. The differences may be caused by influences in the uterus.

Double the Fun: Twins

Identical

There are two types of twins. Identical twins start with one fertilized egg that separates during cell division, making two babies. If the cells don't separate completely, conjoined (Siamese) twins, who share limbs or **organs,** develop. Identical twins have the same **genetic** code because they come from the same egg and sperm!

One Egg

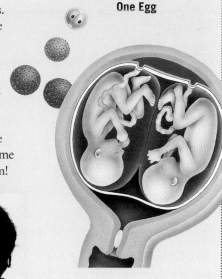

Fraternal

Fraternal twins develop from two different eggs that are released at the same time, then fertilized by two different sperm. Fraternal twins do not have to be of the same sex. They are no more the same in genetic makeup than other siblings. One in 90 U.S. births are twins, and fraternal twins make up 70 to 80 percent of those births.

Two Eggs

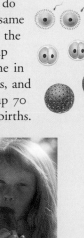

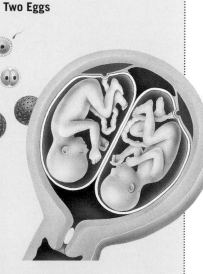

Pregnancy and Birth

Pregnancy is the time between the planting of a **fertilized** egg in the **uterus** *(page 112)* and the birth of a child. It usually lasts 40 weeks, during which a woman's body changes. As the baby grows, it stretches the uterus, and the mother's abdomen grows. So do her breasts, so they can give the baby milk when it is born.

Through the **placenta,** an **organ** joining the mother and baby to be, she nourishes the **fetus** and gets rid of its wastes. The fetus does not breathe for itself until after birth. In the uterus, it floats inside the fluid-filled amniotic sac, which protects it from bumps. The fetus gets all its oxygen through the umbilical cord that connects it to the placenta. This cord is cut after birth. The place where it was attached becomes the baby's bellybutton.

Not all pregnancies go smoothly. Early on, a woman can have a miscarriage (her uterus expels the pregnancy). Some women go into premature labor, and the birthing process starts early. A baby born before six months rarely survives.

Embryo to Fetus

A baby to be is called an **embryo** from the time a fertilized ball of **cells** implants itself into the uterus through the next eight weeks. This is the period when the ground-work for all body structures and organs is laid. The tiny embryo looks more like a tadpole than a baby. From nine weeks through the end of pregnancy, the growing baby is called a fetus. This is a time for rapid growth. Unlike an embryo, a fetus has clearly human limbs and features. Later on, it has a good chance of surviving outside its mother.

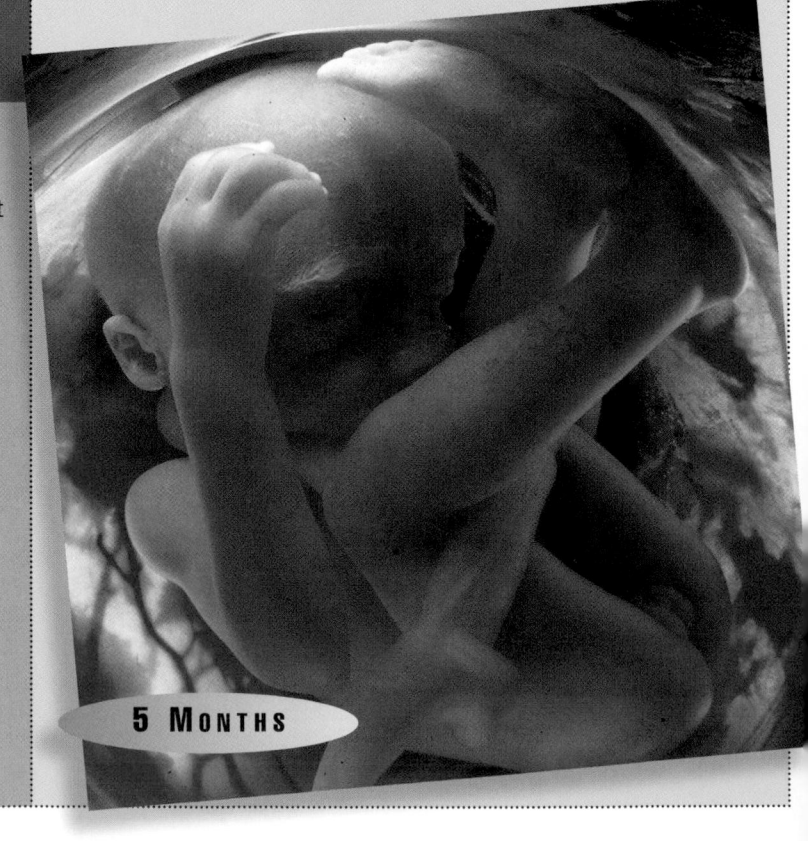

8 WEEKS

5 MONTHS

Storks and Cabbages

There are many legends about where babies come from. For example, in parts of Europe, legend says babies are brought by storks *(left)* from marshes, ponds, wells, or springs. Sometimes people say babies come from cabbage patches. This saying might come from the ancient belief, shared by Greeks, Irish, South Africans, and Indonesians, that trees can give birth to human beings. Old German and Scandinavian stories tell of babies being born first to Mother Earth before coming to human parents.

Nine Months to Grow On

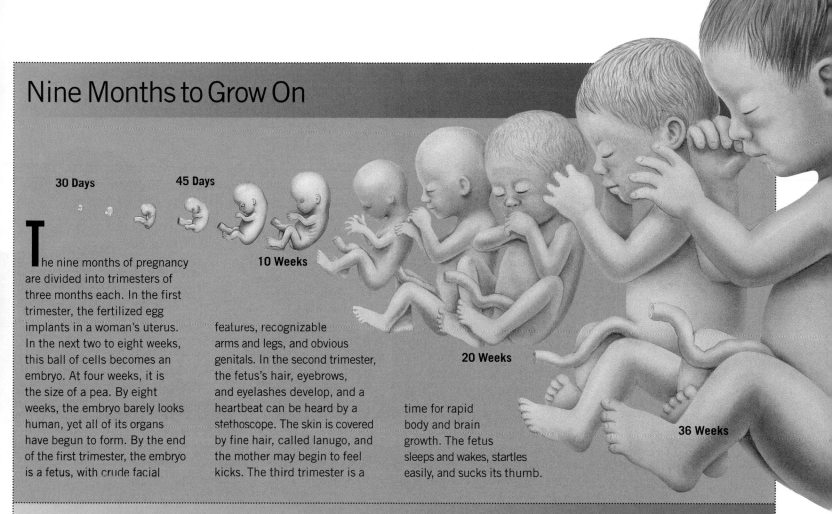

30 Days   **45 Days**

10 Weeks

20 Weeks

36 Weeks

The nine months of pregnancy are divided into trimesters of three months each. In the first trimester, the fertilized egg implants in a woman's uterus. In the next two to eight weeks, this ball of cells becomes an embryo. At four weeks, it is the size of a pea. By eight weeks, the embryo barely looks human, yet all of its organs have begun to form. By the end of the first trimester, the embryo is a fetus, with crude facial features, recognizable arms and legs, and obvious genitals. In the second trimester, the fetus's hair, eyebrows, and eyelashes develop, and a heartbeat can be heard by a stethoscope. The skin is covered by fine hair, called lanugo, and the mother may begin to feel kicks. The third trimester is a time for rapid body and brain growth. The fetus sleeps and wakes, startles easily, and sucks its thumb.

How a Baby Is Born

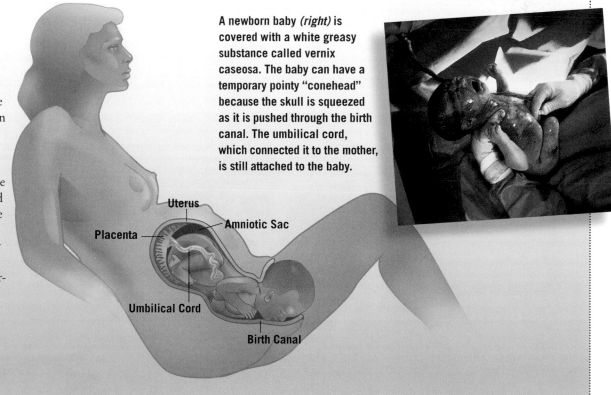

Late in pregnancy, changes in a woman's body signal that the baby is about to be born. Usually, the fetus turns head down and faces backward. This makes it easier to emerge through the birth canal. When the muscular uterus begins to contract, the woman is in labor. The amniotic sac breaks, gushing fluid. In the first stage of labor, the **cervix** widens and thins. During the second stage of labor, the mother squeezes her **muscles,** pushing the baby out. Labor's third stage is the delivery of the **placenta,** or afterbirth. Sometimes, because of risks to the baby or mother's health, a baby is delivered by an operation through the mother's abdomen called a Cesarean section.

A newborn baby *(right)* is covered with a white greasy substance called vernix caseosa. The baby can have a temporary pointy "conehead" because the skull is squeezed as it is pushed through the birth canal. The umbilical cord, which connected it to the mother, is still attached to the baby.

Uterus

Placenta

Amniotic Sac

Umbilical Cord

Birth Canal

What Is Heredity?

Family Resemblance

From the color of your hair to the shape of your body to talents in sports or music, you inherit many **traits** from your parents. These traits are passed along through **genes,** the complicated code that provides the blueprint for building and operating a human being. In fact, every **cell** in your body holds the genes you inherited from your parents. Yet only certain genes are "turned on" in each cell to give that cell its special function.

So far, scientists have found many single genes that control more than 1,500 traits, as well as many more characteristics controlled by polygenes—a number of genes working together. But genes aren't everything. Height, for instance, is controlled by polygenes, but it is also decided by environmental factors like your diet, the amount you sleep, and how much you exercise.

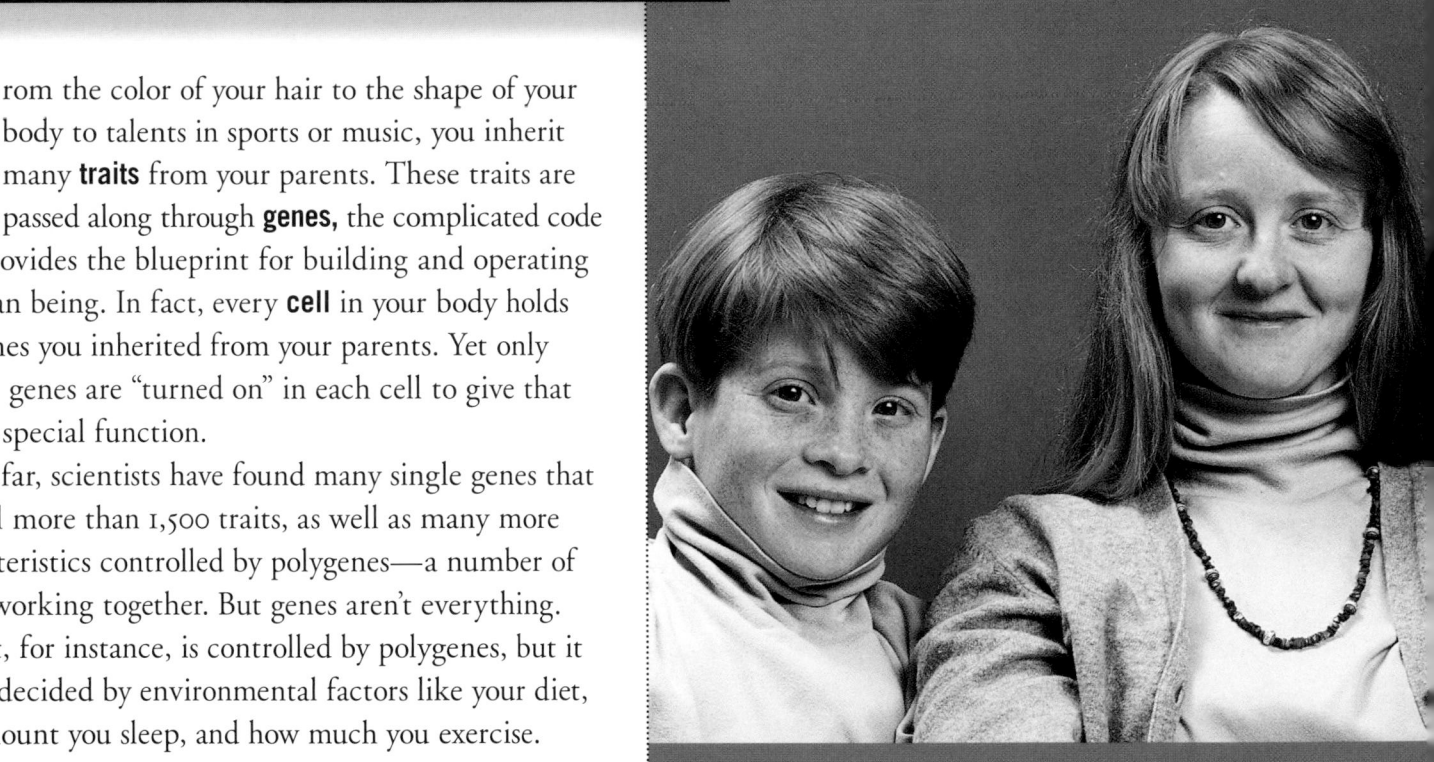

Ever wonder why you have brown eyes like Mom or are tall like Dad? It's called heredity, the characteristics passed from one generation to the next. Family resemblances, like the one seen in this mother and son, show that heredity decides many physical traits.

People — Gregor Mendel

Gregor Mendel, often called the father of modern **genetics,** was born in 1822 in a region of Austria now part of the Czech Republic. After becoming a monk, he went to the University of Vienna, where he learned how to use experiments and math to study nature. In 1856, Mendel began breeding peas in the monastery garden (right) to study **heredity.** Over 10 years, Mendel gathered, counted, labeled, and bred almost 30,000 plants. As he studied his plants he noticed that certain characteristics were carried from one generation to the next. This proved that traits were being inherited—passed from parents to children.

DNA: The Body's Blueprint

The entire blueprint for a human body is inside every cell of the body. These instructions for what the body looks like and how it works are contained in 46 threadlike filaments called **chromosomes,** arranged in 23 pairs (photo at far right). On each pair of chromosomes are thousands of genes. Chromosomes are made of **DNA.** This is short for deoxyribonucleic acid, a **molecule** made up of four types of chemicals called nucleotide bases. These chemicals are arranged as rungs on a long, spiraling ladder called a double helix. If you think of the bases as letters of the alphabet, the pattern in which they are arranged creates gene sentences. A double helix is so tightly coiled that if the DNA in the nucleus of just one human cell were stretched in a straight line, it would reach more than 1.8 m (6 ft.)!

Jeepers Creepers!

Where'd ya get those peepers? Eye color is determined by two forms of a gene at the same place on a pair of chromosomes. These forms, called alleles, are inherited—one from each parent. Although the two alleles are sometimes the same (homozygous), they are often different (heterozygous). The gene for brown eyes (B) is dominant over the gene for blue (b). Children who inherit either a homozygous (BB) or heterozygous (Bb) pair of dominant alleles will be brown eyed, but blue-eyed children can only have homozygous recessive alleles (bb). People with one recessive allele (Bb) are called carriers and can still pass on the trait.

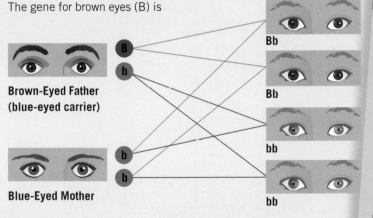

Brown-Eyed Father (blue-eyed carrier)

Blue-Eyed Mother

Bb

Bb

bb

bb

Can You Roll Your Tongue?

If you can, that means that you have a dominant gene for that trait. Not being able to roll your tongue is a recessive trait. Dominant traits always win over recessive traits. Eye color also is determined by dominant and recessive genes *(see feature at left)*. The chart below shows other inherited features and whether they are dominant or recessive.

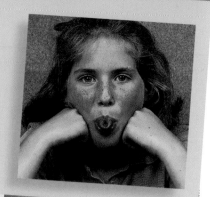

Dominant	Recessive	Dominant	Recessive
Hair		**Eyes**	
Dark	Blond	Brown iris	Blue or gray iris
Curly	Straight	Hazel or green iris	Blue or gray iris
Early baldness *(dominant in males)*	Normal	Large eyes	Small eyes
Widow's peak	Straight hairline	Epicanthic fold	No fold
Skin		**Ears**	
Freckles	Lack of freckles	Free earlobes	Attached earlobes
Black skin	White skin	**Nose**	
Mouth		Roman	Straight
Broad lips	Thin lips	Broad nostrils	Narrow nostrils
Dimples	Dimples absent	**Hands**	
Roll tongue	Unable to roll	Hair on midfinger	Lack of hair
		Pinkie bent in	Pinkie straight

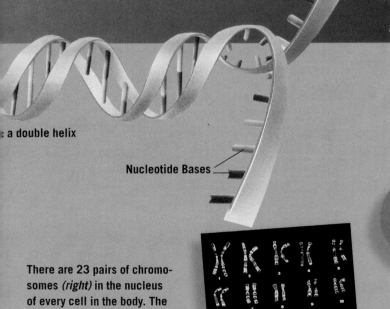

: a double helix

Nucleotide Bases

There are 23 pairs of chromosomes *(right)* in the nucleus of every cell in the body. The 23d pair contains the X and Y chromosomes that determine if a person is a male or a female.

X or Y?

The 23d pair of chromosomes consists of the sex chromosomes. Whereas females have two X chromosomes, males have one X and one Y (which is much smaller than the X). A baby's sex is decided at **fertilization.**

An egg always carries an X chromosome, but a **sperm** carries either an X or a Y. When an X-carrying sperm fertilizes an egg, the baby is a girl (XX); when it's a Y-carrying sperm, it's a boy (XY).

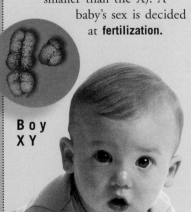

Boy X Y

Girl X X

Growing Up

Human beings take about 20 years to become adults, a period of amazing physical, intellectual, and emotional changes. There are five life stages after birth. During the newborn period—birth through a few weeks—the baby must get used to living outside of the mother's **uterus.** In infancy—from the newborn stage to the end of the first year—a baby learns to focus, grasp, babble, sit, crawl, and finally walk.

Childhood lasts until **puberty,** which is the time of sexual maturation *(opposite).* Childhood is a time of physical growth, skill building, social development, and independent learning.

From sexual maturation through the teen years is adolescence. This is a time when boys and girls differentiate, when girls grow breasts and boys' voices deepen, when the sex **organs** become functional and reproduction is possible. Finally, adulthood is from adolescence through old age. The body is fully mature, but physical aging, as well as intellectual learning and emotional growth, continues.

Body Proportions

This diagram of human growth stages, each drawn to the same height, shows that body proportions change dramatically over time. Before birth, the head is very large compared with the body. A newborn's head is about one-fourth of its total body length. In childhood, the legs grow longer in proportion to the rest of the body, and the child loses baby fat. By adolescence, the head accounts for one-eighth of the body's length.

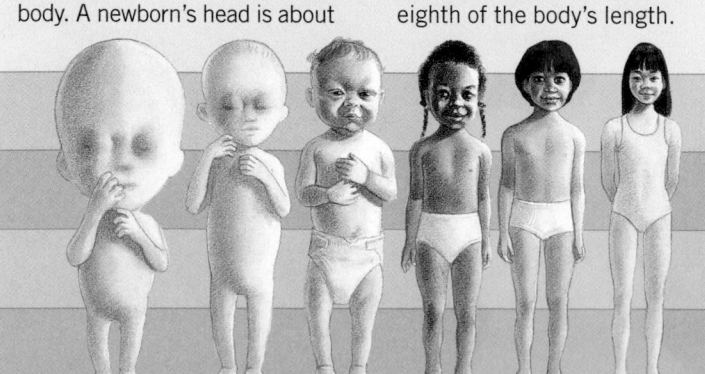

2-month fetus   5-month fetus   Newborn   2 years   5 years   12 years

Rite of Passage

A rite of passage is a symbolic act or ceremony showing that a person is passing from one stage of life into another. Adolescence, the time when a person changes from a child into an adult, is greeted with rites of passage in many cultures. The first period of a traditional Apache girl in the American Southwest sets into motion a yearlong series of events that teach her about the Apache culture and what it means to be an Apache woman. When she is ready, she participates in a four-day coming-of-age ceremony. Starting at sunrise of the first day, she dances for hours around a deerskin-covered blanket to the beat of water drums *(above).* When she finishes dancing, she will lie down on the blanket. Then, an older woman specially chosen to be her spiritual "god-mother" will walk on her back and use her hands to "mold" the girl's body into the woman she has become.

What's Puberty?

Puberty is a time of dramatic physical changes, when the body becomes capable of reproduction. In girls, the average age of puberty is from 10 to 15 years. For girls, puberty is marked with breast budding, a growth spurt, pubic hair, and then the first **menstrual** period. In boys, the average age of puberty is between 11 and 16. It starts with growth of the testes and penis, pubic hair, a growth spurt, the beginnings of facial hair, and deepening of the voice. There is a wide range of "normal" ages and ways to experience puberty.

Why Do We Age?

No one knows for sure why we age, but the cause may be different processes that damage or kill **cells.** Programmed cell death, a natural process, occurs when the brain tells the body's cells to stop dividing. Cells are also damaged or die when they are exposed to environmental toxic chemicals. So your genes and the environment in which you live influence your chances of living a long and healthy life.

Let's Compare

Growth Rates

From birth into the early 20s, young people's bodies change dramatically in size, shape, and proportion. Infancy is the fastest period of growth after birth. In the teenage years, the second fastest period of growth, both boys and girls have a "growth spurt" of 7.6 to 10 cm (3 to 4 in.) between 10 and 14. As you can see from the kids lined up below, boys and girls grow at different rates. At some ages they tend to be about the same height, whereas at others they are taller or shorter. Growth is determined by **genes,** which control **hormone** levels and cell division, as well as by environmental factors like food, exercise, and sleep.

0-2 years   **3-4 years**   **5-7 years**   **8-10 years**   **11-15 years**   **16-20 years**

Picture Credits

Glossary of Terms

Antibody (**an**-ti-bah-dee) A substance produced by the body that can destroy a specific germ, or antigen, by binding to it and that can provide immunity against that specific germ.

Antigen (**an**-ti-jun) A substance, such as bacteria, that activates the immune system.

Aorta (ay-**or**-tuh) The main artery that carries blood from the heart to other arteries.

Arteriole (ar-**tir**-ee-ohl) A small artery.

Artery (**ar**-tur-ee) Blood vessels that carry blood, rich in oxygen, away from the heart.

Atrium (**ay**-tree-uhm) One of the two upper chambers in the heart.

Bacteria (bak-**teer**-ee-uh) A class of microscopic, one-celled or noncellular organisms, some of which are harmful to the body.

Blood pressure (**bluhd presh**-ur) The force of blood pushing against the walls of an artery.

Bone marrow (**bohn mar**-oh) Living, spongy tissue at the center of certain bones that produces blood cells.

Brainstem (**brayn-stem**) Part of the brain that regulates basic functions such as heartbeat and breathing.

Calcium (**kal**-see-uhm) An element necessary for bone hardness; a major component of teeth and bones.

Cancer (**kan**-sur) A disease in which cells reproduce rapidly to form a tumor, an abnormal mass of tissue, which can destroy surrounding healthy cells and can spread to other parts of the body.

Capillary (**kap**-uh-lair-ee) Tiny blood vessels that carry blood from arterioles to venules.

Cartilage (**kar**-tuh-lij) Flexible, elastic tissue that provides cushioning between bones, such as those in the knee and spine.

Cell (**sel**) The basic unit of living organisms.

Cerebellum (ser-uh-**bel**-uhm) The part of the brain that controls skeletal muscle activity such as balance and movement.

Cerebral cortex (suh-**ree**-brul **kor**-teks) The surface layer of the cerebrum; the part of the brain responsible for thinking and learning.

Cerebrum (suh-**ree**-bruhm) The upper and main part of the brain; controls reasoning, memory, and the senses.

Cervix (**sur**-viks) The narrow lower end of the uterus leading to the vagina.

Cholesterol (kuh-**les**-tuh-rol) A substance found in animal fat that can build up and harden on artery walls.

Chromosome (**kroh**-muh-sohm) Structures in the nucleus of a cell, made of DNA, that carry genes that pass on inherited characteristics.

Cilia (**sil**-ee-uh) Microscopic hairlike structures that move like waves.

Circulatory system (**sur**-kyuh-luh-taw-ree **siss**-tuhm) The organs and structures that move blood to and from all parts of the body; the system includes the heart, blood, and blood vessels.

Collagen (**kah**-luh-juhn) A strong, flexible fiber found in the body, especially in bones, connective tissue, and skin.

Conception (verb, **conceive**) (kun-**sep**-shuhn, kun-**seev**) The beginning of pregnancy; the fertilization of an egg by a sperm.

Connective tissue (kuh-**nek**-tiv **tish**-oo) Tissue that supports the body and binds tissues and organs together.

Cranium (**kray**-nee-uhm) The part of the skull that surrounds the brain.

Diaphragm (**dye**-u-fram) A muscle separating the chest cavity from the abdomen. The diaphragm works with chest muscles to draw air into and expel air from the lungs.

Digestive system (dye-**jes**-tiv **siss**-tuhm) The parts of the body that break down food mechanically and chemically so that it can be used by the body.

DNA (**dee en ay**) Deoxyribonucleic acid; molecules in chromosomes that carry hereditary information.

Embryo (**em**-bree-oh) The early stages of development of an organism; in humans the first two months after conception.

Endocrine system (**en**-duh-kruhn **siss**-tuhm) The system of organs and other structures that release hormones.

Enzyme (**en**-zym) Substances produced in the body that aid chemical reactions.

Epidermis (ep-i-**dur**-mus) The outer layer of skin.

Esophagus (i-**sahf**-uh-gus) The muscular tube that leads from the mouth to the stomach.

Fallopian tube (fuh-**loh**-pee-uhn toob) The tube in the female body through which ova move from the ovaries to the uterus.

Fat (**fat**) Tissue filled with greasy or oily material. An important class of energy-rich foods.

Fertilize (**fur**-tuh-lyz) Start of development of a new individual with the joining of an egg and a sperm.

Fetus (**feet**-uhss) A developing human from around the third month of development to birth.

Gene (**jeen**) The basic unit of inheritance. A gene, made of DNA, determines inherited characteristics.

Genetic (juh-**net**-ik) The study of the way characteristics are passed on from parents to offspring.

Genital (**jen**-uh-tuhl) The external sex organs.

Germ (**jurm**) A microscopic organism, such as bacteria and viruses, that causes disease.

Gland (**gland**) A part of the body that removes certain materials from the blood and concentrates or alters them for further use in the body or for elimination.

Glucose (**gloo**-kohss) A sugar that is formed by the digestion of carbohydrates, glucose provides energy for the body and is the main sugar in the blood.

Heredity (huh-**red**-i-tee) The passing of characteristics, such as hair color, from parents to offspring through genes.

Hormones (**hor**-mohnz) Chemical messengers released into the bloodstream by certain glands that regulate and keep body processes in balance by changing behavior of cells in another part of the body in a specific way.

Hypothalamus (hy-puh-**thal**-uh-muhss) The part of the brain that controls automatic functions such as temperature and heartbeat.

Immune system (im-**yoon** siss-tuhm) Organs and structures in the body that help fight infectious diseases.

Immunity (im-**yoo**-nuh-tee) The ability of the body to resist substances that can cause disease.

Joint (**joynt**) A place where bones connect.

Ligament (**lig**-uh-muhnt) Tough, ropy tissue that connects bones.

Lymph (**limf**) A fluid carrying white blood cells, similar to blood plasma, that circulates in vessels that serve as part of the immune system.

Macrophage (**mak**-ruh-fayj) A white blood cell that consumes germs and damaged cells.

Membrane (**mem**-brayn) A thin, soft covering of tissue.

Menstruation (men-**stray**-shuhn) The monthly shedding of the uterine lining in nonpregnant sexually mature females.

Metabolism (muh-**tab**-uh-liz-uhm) All the chemical and physical processes that occur in the body; the chemical processes in cells in which new substances are taken in and energy is produced.

Mineral (**min**-ur-uhl) Elements or chemical compounds that are needed by the human body—for example, calcium and iron.

Molecule (**mahl**-uh-kyool) A particle containing one or more atoms.

Mucus (**myoo**-kuhss) A thick, protective liquid that lines the respiratory and digestive systems.

Muscle (**mus**-uhl) Tissue that contracts, or shortens, when it is stimulated and as a result produces motion.

Nerve (**nurv**) Fibers that send messages both to and from the brain and the spinal cord.

Nervous system (**nur**-vuhss siss-tuhm) The organs and structures of the body, including the brain, the spinal cord, and nerves, that receive and interpret sensations and transmit impulses.

Neuron (**noo**-rahn) Nerve cell.

Nucleus (**noo**-klee-uhss) The area of a cell, usually in the center, that contains the cell's chromosomes.

Nutrient (**noo**-tree-uhnt) Microscopic substances that are produced by digested food and are used by cells for energy, growth, and repair.

Optic nerve (**ahp**-tik **nurv**) The main nerve linking the eye to the brain.

Organ (**or**-guhn) A group of tissues, such as the heart, kidney, and eyes, that work together to perform a specific job.

Ovary (**oh**-vuh-ree) The female organ in which ova, or eggs, are produced and stored.

Ovulate (**oh**-vyuh-layt) The release of an egg by the ovary.

Ovum (plural **ova**) (**oh**-vuhm, plural **oh**-vuh) The female reproductive cell, or egg. When joined with a sperm it can develop into a new individual.

Pelvis (**pel**-vuhss) A basin-shaped structure formed by the hipbones and backbone.

Pigment (**pig**-muhnt) Matter that gives color in a cell or tissue.

Pituitary gland (puh-**too**-uh-ter-ee gland) A gland in the brain that controls other hormone-producing glands and helps regulate a variety of functions including growth and sexual development.

Placenta (pluh-**sent**-uh) The organ that connects a fetus in the process of developing to its mother. The placenta provides nourishment and removes wastes.

Plasma (**plaz**-muh) The liquid part of blood in which blood cells float.

Protein (**proh**-teen) A class of molecules with a wide variety of types and functions; the main component of cells. Protein, an important part of the human diet, is used for growth and repair.

Puberty (**pyoo**-bur-tee) The period of human development during which reproductive organs become functional; sexual maturation.

Pulse (**puls**) The rhythmic throbbing heartbeat that is felt in arteries close to the surface of the skin.

Pupil (**pyoo**-puhl) The opening in the iris through which light enters the eye.

Receptor (ri-**sep**-tur) A nerve ending that is capable of detecting a particular type of sensation such as pain, heat and cold, or pressure.

Glossary of Terms

Reflex (**ree**-fleks) An automatic response made without thinking.

Reproductive system (ree-pruh-**duk**-tiv **siss**-tuhm) The organs and other structures in the male and female body responsible for production of ova or sperm, fertilization, development, and the birth of a baby.

Respiratory system (res-pruh-taw- ree **siss**-tuhm) The organs and structures involved in breathing, including the lungs and trachea.

Saliva (suh-**lye**-vuh) Liquid produced in the mouth by salivary glands that softens and begins the digestion of food.

Sperm (**spurm**) Reproductive cell in males that when joined with an ovum, or egg, can develop into a new individual.

Spinal cord (**spye**-nuhl **kord**) The bundle of nerves running from the base of the brain down through the vertebrae in the spine.

Spleen (**spleen**) An organ near the stomach that produces some types of white blood cells and removes dead red blood cells and platelets from the blood.

Sternum (**stur**-nuhm) The bone in the front of the chest to which most of the ribs are attached; breastbone.

System (**siss**-tuhm) A group of organs that work together to perform a particular job. There are 11 major systems in the body, including the circulatory system and digestive system.

Tendon (**ten**-duhn) Tough, ropy tissue that connects muscle to bone.

Tissue (**tish**-oo) A group of cells that are similar in their composition and that perform a specific job.

Trait (**trayt**) A characteristic that is inherited.

Uterus (**yoot**-uh-russ) The organ in the female body where a baby grows and develops before birth. The uterus is also known as the womb.

Vaccine (**vak**-seen) Dead or weakened antigens prepared in a form that can be injected or swallowed. A vaccine gives a person immunity against a disease by building up antibodies in the blood.

Vein (**vayn**) Blood vessels that carry blood low in oxygen back to the heart.

Ventricle (**ven**-tri-kuhl) One of the two lower chambers of the heart that receive blood from the atrium. On the right side the ventricle sends blood to the pulmonary artery, and on the left it sends blood to the aorta.

Venule (**ven**-yool) A small vein.

Virus (**vye**-ruhss) Matter that can cause disease by multiplying within living cells.

Vitamins (**vye**-tuh-minz) Substances that are required by the body in small amounts in order for growth and maintenance to take place.

Index

Index

Time-Life Education, Inc. is a division of Time Life Inc.

TIME LIFE INC.

PRESIDENT and CEO: George Artandi
CHIEF OPERATING OFFICER: Mary Davis Holt

TIME-LIFE EDUCATION, INC.
PRESIDENT: Mary Davis Holt
MANAGING EDITOR: Mary J. Wright

Time-Life Student Library
HUMAN BODY

SERIES EDITOR: Jean Burke Crawford

Associate Editor/Research and Writing: Mary Saxton
Series Picture Associate: Lisa Groseclose
Editorial Assistant: Maria Washington
Picture Coordinator: Daryl Beard

Designed by: Jeff McKay and Phillip Unetic, 3r1 Group

Special Contributors: Janet Cave, Patricia Daniels, Susan Perry, Terrell Smith, Marilyn Terrell (text); Susan Blair, Patti Cass (research); Barbara Klein (index).
Senior Copyeditor: Judith Klein
Correspondents: Maria Vincenza Aloisi (Paris), Christine Hinze (London), Christina Lieberman (New York).
Editorial Interns: Renesa Bell, Lenese Stephens, Tonya Wilson.

Vice President of Marketing and Publisher: Rosalyn McPherson Andrews
Vice President of Sales: Robert F. Sheridan
Director of Book Production: Patricia Pascale
Director of Publishing Technology: Betsi McGrath
Director of Photography and Research: John Conrad Weiser
Marketing Manager: Michelle Stegmaier
Production Manager: Carolyn Clark
Quality Assurance Manager: James King
Chief Librarian: Louise D. Forstall
Direct Marketing Consultant: Barbara Erlandson

Consultants: Caroline Wellbery, M.D., is an assistant professor in the Department of Family Medicine at Georgetown University in Washington, D.C. She is also assistant deputy editor of the journal *American Family Physician.* She writes extensively on medically related subjects for newspapers and medical journals.

Lisa Lyle Wu teaches biology, marine biology, and avanced-placement biology at Thomas Jefferson High School for Science and Technology in Fairfax County, Virginia, and has taught biology since 1978. She is an education specialist for the Smithsonian Institution's National Museum of Natural History and has developed curriculum materials for the Discovery Channel and the National Audubon Society.

Library of Congress Cataloging-in-Publication Data
Human body.
        p. cm. — (Time-Life student library)
        Includes index.
        Summary: Examines the structure and function of various parts of the human body, including bones, muscles, heart, lungs, brain, nervous system, digestive system, immune system, and reproductive organs.
        ISBN 0-7835-1353-4
        1. Body, Human—Juvenile literature. [1. Body, Human. 2. Human physiology. 3. Human anatomy.] I. Time-Life Books. II. Series.
        QP37.H89 1999
612—dc21   98-53020
                                                                            CIP
                                                                            AC

OTHER PUBLICATIONS

TIME-LIFE KIDS
Library of First Questions and
    Answers
A Child's First Library of Learning
I Love Math
Nature Company Discoveries
Understanding Science & Nature

HISTORY
Our American Century
World War II
What Life Was Like
The American Story
Voices of the Civil War
The American Indians
Lost Civilizations
Mysteries of the Unknown
Time Frame
The Civil War
Cultural Atlas

SCIENCE/NATURE
Voyage Through the Universe

DO IT YOURSELF
Total Golf
How to Fix It
The Time-Life Complete Gardener
Home Repair and Improvement
The Art of Woodworking

COOKING
Weight Watchers® Smart Choice
    Recipe Collection
Great Taste–Low Fat
Williams-Sonoma Kitchen Library

For information on and a full description of any of the Time-Life Books series listed above, please call 1-800-621-7026 or write:

Reader Information
Time-Life Customer Service
P.O. Box C-32068
Richmond, Virginia 23261-2068